Animal Welfare

A Concise Guide for Veterinary Professionals

Animal Welfare

A Concise Guide for Veterinary Professionals

Kathy Anzuino, Joe Anzuino and Paul Roger

CABI

CABI is a trading name of CAB International

CABI
Nosworthy Way
Wallingford
Oxfordshire OX10 8DE
UK

Tel: +44 (0)1491 832111
E-mail: info@cabi.org
Website: www.cabi.org

CABI
200 Portland Street
Boston
MA 02114
USA

Tel: +1 (617)682-9015
E-mail: cabi-nao@cabi.org

A catalogue record for this book is available from the British Library, London, UK.

ISBN-13: 9781786391568 (hardback)
9781786391551 (paperback)
9781786391575 (ePDF)
9781786391582 (ePub)

DOI: 10.1079/9781786391582.0000

Commissioning Editor: Alex Lainsbury
Editorial Assistant: Theresa Regueira
Production Editor: James Bishop

Typeset by Straive, Pondicherry, India
Printed and bound by CPI Group (UK) Ltd, Croydon, CR0 4YY

Contents

Foreword

I was extremely pleased to be asked to contribute to this book, as I have been involved in the subject's uptake by the veterinary profession for many years. Before I do, I thought it might be of interest to give some of my insights into the subject's history and evolution since the mid-1960s, particularly regarding the veterinary profession.

Animal welfare has been a concern of society for over a hundred years. In the UK, legislation dates back to the nineteenth century, with the Protection of Animals Act in 1841, commonly known as the 'prevention of cruelty to animals act'. Later publications, such as Animal Machines (1964) by Ruth Harrison, further raised public awareness of animal welfare on so-called factory farms.[1] However, views on animal welfare were largely based on anthropomorphic feelings. The increasing emphasis on the role of science in society stimulated a search for evidence to support these strong feelings. This search began in the UK but was later extended to many other countries, and has been important for the development of animal welfare science.

In 1965, the UK government commissioned the Brambell Report on the welfare of farmed animals.[2] The report proposed an ethical framework that later became internationally adopted and known as 'The Five Freedoms'. Appendix III in the report - the 'Assessment of Pain and Distress in Animals' by W. H. Thorpe - was perhaps the most prescient as it started to map out ways in which pain and distress could be separated, recognised and evaluated using physiological and ethological (behavioural) indicators. Appendices IV and V also provided valuable theoretical insights into animal suffering, introducing the concept of 'mental suffering' in animals, or 'mental health' as it is often interpreted nowadays. The Report also recommended that a Statutory Farm Animal Welfare Standing Advisory Committee be established to address on-going concerns, which later became the Farm Animal Welfare Council (FAWC 1979-2011); then the Farm Animal Welfare Committee (FAWC 2011-2019), and more recently the Animal Welfare Committee (AWC 2019-present). FAWC began laying down standards for the welfare of farmed animals, with practical guidelines set out in Codes of Practice that, in turn, led to further research. However, sentience and suffering in other animal species were largely put aside. For example, millions of other species were being used in painful procedures in experimental animal research and subjected to major challenges to their welfare but their use for a scientific purpose exempted them from regulation providing protections for farmed animals.

Of particular concern for experimental animals was the severity of the scientific procedures being carried out. In 1985, Paul Griffiths and I wrote a paper (both of us were laboratory animal veterinarians at the time) on the recognition of pain, distress and discomfort in experimental animals[3] as an aid in updating the 1876 Act. Subsequently, this law was revised and became the Animals (Scientific Procedures) Act 1986. The idea for the paper came from Paul, whose dog was involved in a road accident, and he was very concerned about how he could recognise whether his dog was in pain. In our paper, I added the 'how to measure' hypothesis. Interestingly, in those days, i.e., up to the late 1980s, veterinary professionals treating companion animals rarely considered providing analgesia, pre- or post-operatively, for their patients. Certainly, analgesia was very limited in farmed animals, despite the numerous mutilations they endured, such as castration, dehorning, disbudding, and docking.

In the 1990s, the veterinary profession began to take animals' mental health, alongside their physical health, more seriously. This shift was stimulated by various groups 'taking the lead', such as antivivisection societies, animal welfare charities, and biologists such as ethologists.

The profession could, and should, be an important voice in the development of animal welfare, along with the other groups. For example, veterinary professionals had the educational background and access to the necessary drugs, particularly for anaesthesia and analgesia. They also knew that some diseases could exacerbate pain and distress in animals and how to recognise key signs. As a consequence, the Royal College of Veterinary Surgeons (RCVS) started a post-graduate qualification for veterinarians. I was involved in its setting up and drawing up the initial syllabus, emphasising its scientific basis – after all, we were qualified in Veterinary 'Science' – which again helped to persuade the RCVS to recognise the speciality.

This persuasion was needed as most veterinarians at that time felt strongly that 'animal welfare' was what they did on a daily basis, but did not fully appreciate that animal welfare involves full consideration of both the physical health (free from infectious and non-infectious disease) and the mental health of animals (e.g. free from fear, anxiety, pain and distress), and that both could be underpinned by credible evidence.

The resistance of the RCVS was overcome, and the Certificate in Animal Welfare Science, Ethics and Law (AWSEL), otherwise known as CertWEL, was established in 1995. The logic behind linking this trio of subjects was that once a welfare problem was identified and backed up by a sound scientific basis, there was an ethical imperative to do something about it and to avoid causing animal suffering; this, in turn, could, and did, lead to changes in the law. The first graduates of the Certificate in Animal Welfare Science, Ethics and Law

(AWSEL) in 1996 were Ruth Layton, Paul Roger, and Roger Hancock, followed by Ed Varley, Heather Elliott, Caroline Manser, Kate Kerr, and Sue Haslam. By 2011, there were 48 graduates.

In 1999, the Diploma in AWSEL (DWEL) was established, providing for a deeper and broader knowledge of animal welfare than the Certificate did. The same syllabus and examination format would be followed, but the original examiners were now replaced by those graduating with the CertWEL. The Diploma was later subsumed by the European Board of Veterinary Specialisation (EBVS), with the establishment of the European College of Animal Welfare and Behavioural Medicine (ECAWBM). In 2012, there were 10 vets in the UK holding the DWEL, and there are now, internationally, well over 100 with the Diploma in AWSEL.

=========

Now back to the book! I have been involved in promoting this trio of subjects at the academic and educational levels, and it is gratifying to see this publication from my former students.

The authors primarily aim their work at veterinary professionals and, drawing on their experience working in teams in the UK and internationally, they have defined veterinary professionals as including not only veterinarians and veterinary nurses but also other members of the veterinary workforce who contribute to safeguarding and improving animal welfare. This would also be an ideal book for anyone interested in animal welfare, such as agricultural students, students studying animal welfare and related subjects such as ethology and physiology, as well as members of animal welfare charities/non-governmental organisations. Moreover, the book will be of interest to those who wish to engage further with the ethical and legal issues that are raised and where practical knowledge and facts are needed.

The welfare of all domesticated and wild animals can be seen as a public good that must be taken seriously by politicians and society. However, while that is a laudable aim, achieving it is fraught with difficulty. This book provides guidance by providing basic frameworks for animal welfare, ethics and law in its first half and then applies these to several of the common domesticated, farmed and companion animal species in the second part, as well as including two other animal groups – laboratory and wild. Linking the three strands of the subject matter provides a broad critique of how we use animals today.

All three authors are uniquely qualified to write a book in this area, and are supported by many colleagues who have provided, as they put it, 'additional insights and feedback on various chapters'. They combine experience across different fields of veterinary endeavour, including general practice, academia, charities/non-governmental organisations, and government, with a passion for advancing understanding and promoting animal welfare. They introduce these topics using their wealth of experience and practical knowledge, taking them beyond theory into real-life, practical application.

As they state in their introduction: '*By reading this book, veterinary professionals should be in a stronger position to advocate for animals. They should be better placed to analyse welfare holistically from first principles, whatever the species or context, and be more aware of the welfare issues facing all animals commonly used by society, not just those they work with directly*'.

I believe that readers will learn from this book and enjoy its content.

David Morton
CBE BVSc PhD FSB DipECLAM DipECAWBM MRCVS
Professor emeritus (University of Birmingham)

---

[1] Harrison, R., 1964. Animal machines: the new factory farming industry. Vincent Stuart Ltd, 45 Lower Belgrave Street, London SW1. 186 pp.

[2] Brambell Committee, 1965. Report of the technical committee to enquire into the welfare of animals kept under intensive livestock husbandry systems. Command Report 2836, Her Majesty's Stationery Office, London.

[3] Morton, D.B. & Griffiths P. H. M. (1985) Guidelines on the Recognition of Pain, Distress and Discomfort in Experimental Animals and an Hypothesis for Assessment. Veterinary Record 116, 431–436.

About this book

This book is based on the premise that veterinary professionals can, and should, be leading advocates for animal welfare, with action underpinned by in-depth knowledge, integrity and compassion. It does not assume that veterinary professionals will ignore factors outside of animal welfare; indeed, it is the pragmatism and multi-disciplinary skill sets of veterinary professionals that enable them to work effectively with many different stakeholders to bring about change. However, it does mean ensuring that when decisions about animals are made, their interests are included, properly represented and respected, a concept we refer to as advocacy. It includes striving for animal welfare to become mainstream in society, prioritised at least to the same extent as topical issues such as environmental protection and public health.

In this book, the term veterinary professionals includes veterinarians and veterinary nurses, whether qualified or students, working in diverse areas such as clinical practice, teaching, research and industry. It also covers veterinary paraprofessionals, who are skilled and valued members of veterinary teams. Veterinary professionals are ideally placed to lead on animal welfare through their possessing a strong science-based knowledge about animal health and husbandry, good technical skills to apply this information practically, empathy for animals and a desire to protect them, alongside regular interactions with animal carers and other stakeholders who impact animal welfare, such as scientists, policymakers, industry and charities. In many communities, they have long-standing credibility through adhering to high ethical and professional standards.[1] However, while veterinary professionals already cater well for some aspects of animal welfare, others are overlooked, and much more can be done.

Animal welfare is now a well-established discipline, having developed substantially over the last thirty years. Key individuals – Mike Appleby, Donald Broom, Marian Dawkins, David Fraser, Temple Grandin, David Morton and John Webster, to name a few – have shaped our understanding and sense of how to advocate for animals in our veterinary work, for which we owe them an enormous debt of gratitude. It is positive that there is now a huge amount of information and evidence base that veterinary professionals can engage with.

However, the downside is that the sheer amount of information can be overwhelming, making it difficult to 'see the wood for the trees'. Therefore, we aim to make the subject area much easier for our busy, time-strapped colleagues to navigate, helping them 'fast-track' to the knowledge the contributors to the book have developed gradually over many years through specific training and practical experience.

The book is written in a 'plain English' style, emphasising clarity of concepts over nuance in order to provide readers with a solid foundation for thinking about animal welfare. It is divided into two main parts. Part 1 introduces key concepts in animal welfare, ethics and the law. Part 2 builds on Part 1 by illustrating how key welfare concepts apply to different species and contexts. It does not aim to provide encyclopaedic knowledge for each species. Instead, the aim is to illustrate a way of thinking about welfare that helps readers engage with more complex sources and the inevitable real-world nuance.

Part 2 chapters generally follow a similar format. The species and their common usages are introduced. The context that will be discussed in further detail is specified, having been narrowed to strike a helpful balance between depth and breadth. Sufficient detail on husbandry and production systems is provided to help veterinary professionals engage with the topic. For instance, many veterinary professionals will already be familiar with the way many owners keep companion dogs and cats, but will likely be much less knowledgeable about husbandry systems for farmed pigs, chickens, rabbits and fish, unless they specialise in these areas, so the latter requires considerably more explanation. Common welfare hazards (quality of the environment, feed, management and so on) are listed. Common welfare outcomes are then identified and structured according to the Five Freedoms framework. Practical welfare assessment protocols available at the time of writing and potentially useful for veterinary professionals are referenced. As practical action is needed to improve welfare, examples of how veterinary professionals have or can become involved in advocacy at the local, national or international levels are given. Further credible resources that we have found helpful are signposted. While full ethical analyses are outside the scope of this book, it is essential to remember that ethics and welfare are intertwined. Further ethical discussions are key in deciding acceptable levels of welfare and whether certain animal uses can be justified, for which Chapter 4 provides further guidance. Law is a very different discipline from veterinary medicine, for which Chapter 5 provides a basic introduction.

By reading this book, veterinary professionals should be in a stronger position to advocate for animals. They should be better placed to analyse welfare holistically from first principles, whatever the species or context, and be more aware of the welfare issues facing all animals commonly used by society, not just those they work with directly. Hopefully, they will ask more probing ethical questions and lead by example in their personal choices.

Contributors

The following people have generously contributed their knowledge and experience to this book:

Jo White BA Equine Studies, MSc Human Behaviour Change, Cert Campaigns of Human Behaviour Change for Life CIC (Chapter 2);
Kate Still, Soil Association (Chapter 3);
Christine Leeb Assoc. Prof. PD Dr. med.vet. Dipl ECAWBM (AWSEL) (Chapter 6);
Richard Jennison BVM&S, CertLAS, CertWEL, Cert PMP, MRCVS (Chapter 7);
Katrine Bazeley BSc, BVSc, CertWEL (Chapter 8);
Sean Wensley BVSc, MSc, FRCVS (Chapter 12);
Madeleine Campbell BVetMed (Hons), MA (Oxon), MA (Keele), PGCertVetEd, PhD, DipECA, DipECAWBM (AWSEL), FHEA, FRCVS, European Veterinary Specialist and RCVS Specialist in Animal Welfare Science, Ethics and Law (Chapter 13);
Chris Walster BVM&S, MVPH, CertAqV, MRCVS (Chapter 14);
Matthijs Metseelar DVM, PhD, Dipl ECAAH, CertAqV, MIFM, MRCVS Specialist Aquatic Veterinary Surgeon, RCVS Specialist in Fish Welfare and Production (Chapter 14);
Heather Bacon OBE, BVSc, Cert ZooMed, PhD, MRCVS (Chapter 15);
Sarah Wolfensohn OBE BSc MA VetMB FRSB DipECAWBM FRCVS Emerita Professor of Animal Welfare (Chapter 16);
Marianne Farish, Humane Slaughter Association (Chapter 17).

We want to express our sincere thanks to friends, professional colleagues, and teachers not named above whose guidance, support, and insights have contributed to the development of this work.

While grounded in the experiences of British veterinary professionals, we hope this book will be valuable to readers in other countries and cultures, helping them adapt advocacy approaches to their local contexts.

Image Credits:
Figure 2-2: U. Rosenschild / iStock; 2.3: Wirestock / iStock; 4.1 (right): N. Somsavangwong / iStock; 4.2: Pixelfusion3d / iStock; 6.2: Y. Srimongkol /

iStock; 6.3: Acceptfoto / iStock; 6.4: Daisy-Daisy / istock; 6.5: C. Lotongkum / iStock; 6.6: A. Brunsting / iStock; 7.1: Rittaewa / iStock; 7.2: JackF / iStock; 7.3: P. M. Faus / iStock; 7.4a: L. Woods / iStock; 7.4b: M. Hobson / iStock (Free Range); 7.5: A. Ribeiro / iStock; 8.1: T. Sylapakith / iStock; 8.2: Ballgally View Images / iStock; 8.3: P. Kstanisz / iStock; 8.4: Friend2727 / iStock; 8.5: Bierwirm / iStock; 9.1: N. Burley / iStock; 9.2: R. Gosch / iStock; 9.3: J. P. Hutchinson / iStock; 10.1: A. Segre / iStock; 12.1: KatuSka / iStock; 12.2: Zeleno / iStock; 13.1: R. Olinchuk / iStock; 13.2: EyeEm Mobile / iStock; 13.3: VectorMine / iStock; 14.1: GordonImages / iStock; 14.2: Marius.tu / iStock; 15.1: I. Barnet / iStock; 15.2: H. Pechan / iStock; 15.3: K. Mattakulphon / iStock; 16.1: Unol / iStock.

Graphic input assistance: Ashram Ashraf

Figure 5.1: The Trial of Bill Burn Under Martin's Act, Print by Charles Hunt after original by P. Matthews, © Parliamentary Art Collection, WOA 6299. Printed with Permission.

We extend our sincere thanks to World Animal Protection (www.worldanimalprotection.org) for supporting the development of the STEPS video and for granting permission to share these videos via the CABI platform (Chapter 17).

This book is dedicated to two dear veterinary colleagues who are sadly no longer with us, Sue Haslam and David Bayvel, both stalwarts of animal welfare and advocacy.

Part 1

Animal Welfare Concepts

1

Abstract
Key concepts and frameworks for holistically assessing an animal's welfare are discussed. The importance of physical and mental health, and their relationship with the naturalness of an animal's life, is explained. Several welfare assessment frameworks are overviewed: the Five Freedoms, the Five Domains Model, Welfare Quality® and the Five Welfare Needs.

Introduction

The many animals that humans commonly interact with are sentient, meaning they can feel a range of sensations and emotions, both pleasant and unpleasant, and as such, they are of moral concern.

Ultimately, welfare concerns how animals feel. Another term commonly used is 'quality of life'. At its most basic, assessing animal welfare involves asking the following two questions:

1. Are the animals healthy?
2. Do they have what they want?[2]

Physical health is obviously of fundamental importance for animal welfare. However, animals have needs and desires in addition to good physical health – they need good mental health. As veterinary professionals, our understanding of animals' social and other behavioural needs has often lagged behind our understanding of physical health, and we have sometimes been criticised for ignoring these aspects. However, such information is increasingly available and should be part of our evidence base.

DOI: 10.1079/9781786391582.0001

Animal welfare frameworks

Frameworks provide a structure for thinking about welfare. They act as prompts to ensure that all areas important for welfare are considered. Four frameworks commonly used are:

1. The Five Freedoms.
2. The Five Domains Model.
3. Welfare Quality®.
4. The Five Welfare Needs.

The Five Freedoms

Most veterinary professionals will be familiar with the Five Freedoms. It is internationally recognised and probably the most widely used framework to date. Ruth Harrison's seminal book titled 'Animal Machines',[3] published in 1964, triggered much public interest in intensive farming in the UK when it was serialised in a popular newspaper. This led to an official enquiry into the welfare of animals kept under intensive livestock husbandry systems, published as the Brambell Report in 1965.[4] This report stated that farm animals should have the freedom 'to stand up, lie down, turn around, groom themselves and stretch their limbs'. The British Government then established the Farm Animal Welfare Advisory Committee, which was replaced in July 1979 by the Farm Animal Welfare Council (FAWC), now known as the Animal Welfare Committee (AWC).[5] (This is a specialised welfare group, mentioned throughout this book in relevant sections, as they have produced and continue to release excellent reports covering various welfare issues.) The exact origins of the Five Freedoms are unclear. One of the early bodies listed provisions important for farm animal welfare, placing criteria in five broad categories, which became known as the 'Five Freedoms', despite not all categories being 'freedoms'. The earliest record appears to be the press notice released by FAWC in December 1979 specifying that any Welfare Codes for livestock should be revised to provide animals with freedom from thirst, hunger and malnutrition; appropriate comfort and shelter; prevention or rapid diagnosis and treatment of injury and disease; freedom to display most normal patterns of behaviour; and freedom from fear. FAWC subsequently refined these to produce the 'Five Freedoms' known today, shown in bold below, along with the additional caveats that explain how they might be met. (We have labelled them F1 through F5 for clarity when used in Part 2 of this book.)

F1 **Freedom from hunger, thirst and malnutrition** by ready access to fresh water and a diet to maintain full health and vigour.
F2 **Freedom from discomfort** by providing a suitable environment, including shelter and a comfortable resting place.

F3 Freedom from pain, injury and disease by prevention, rapid treatment and diagnosis.
F4 Freedom from fear and distress by ensuring conditions which avoid mental suffering.
F5 Freedom to express normal behaviour by providing sufficient space, proper facilities and company of an animal's own kind.

The Five Freedoms prompt us to consider all areas important for welfare, and not just focus on one or two of them. **A key caveat when using the Five Freedoms framework is that positive experiences must be considered, not just freedom from unpleasant experiences.**

The Freedoms are interrelated. Also, they are not absolute and must be interpreted with common sense. Intensity and duration are important. For example, an animal must feel at least some hunger to stimulate it to eat, but for welfare reasons, prolonged hunger and thirst must be avoided. The details of what each freedom means for different species in different situations must be carefully interpreted, using the available scientific evidence and broad experience gained working with animals. For example, consider 'Freedom to express normal behaviour'. It is essential to know which species are solitary and which are social. It is important to know which normal behaviours the animal wants to do, under what circumstances, and which it prefers to avoid. John Webster, who was instrumental in developing the Five Freedoms, suggests that 'Freedom to express normal behaviour' may read better as 'Freedom of choice',[6] as welfare is often improved when animals have some control over their situation.

As the academic debate on welfare continues, the content of the Five Freedoms continues to be represented in slightly different ways by different authors. The Five Domains, the Welfare Quality® framework and the Five Welfare Needs have evolved from the Five Freedoms.

The Five Domains Model

The Five Domains model[7] (Figure 1-1) covers similar areas as the Five Freedoms, but further emphasises that it is the animal's feelings that matter.

Welfare Quality®

The Welfare Quality® framework covers the same areas as the Five Freedoms but emphasises that providing animals with positive life experiences, not just avoiding negative ones, is important for their welfare. The welfare principles (see Table 1-1) use the terms 'good' and 'appropriate'.

This framework has been extensively used in developing welfare assessment protocols within the European Union (EU) under the umbrella of the Welfare Quality® project.[8]

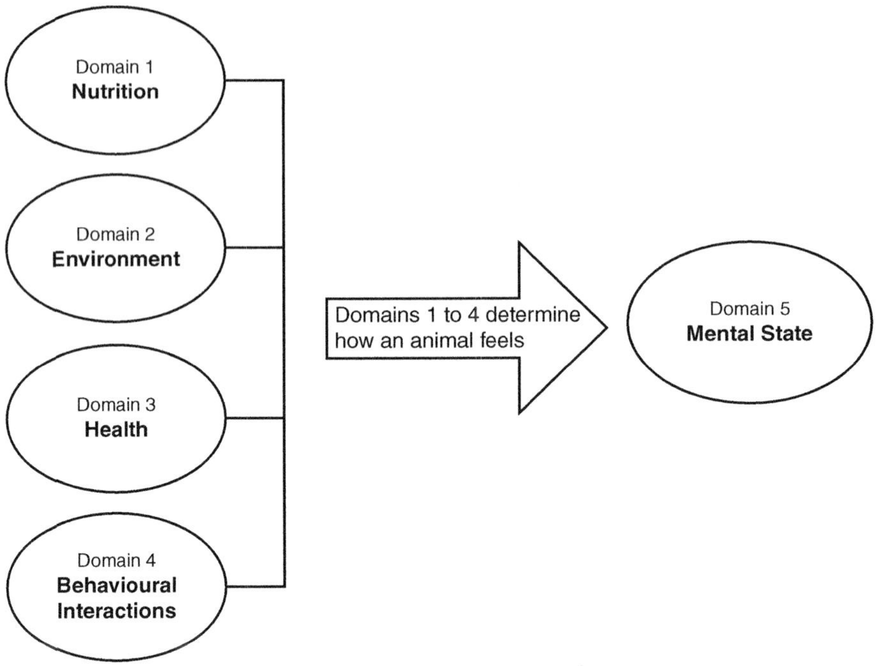

Figure 1-1. The Five Domains.

Table 1-1. The welfare principles and criteria comprising the Welfare Quality® framework.

Welfare principles	Welfare criteria
1. Good feeding	**1.** Absence of prolonged hunger
	2. Absence of prolonged thirst
2. Good housing	**3.** Comfort around resting
	4. Thermal comfort
	5. Ease of movement
3. Good health	**6.** Absence of injuries
	7. Absence of disease
	8. Absence of pain-induced management procedures
4. Appropriate behaviour	**9.** Expression of social behaviours
	10. Expression of other behaviours
	11. Good human-animal relationships
	12. Positive emotional state

The Five Welfare Needs

The Animal Welfare Act (AWA) 2006 replaced outdated animal protection laws in England and Wales, assigning owners and carers of animals a legal duty of care. Owners and carers must provide for their animals' needs, which were determined using the Five Freedoms framework and then translated into suitable wording for the legal context (section 9 of the AWA 2006). These have become known as the Five Welfare Needs and are frequently used as a framework for welfare assessments. They are:

- The need for a suitable environment.
- The need for a suitable diet.
- The need to be able to exhibit normal behaviour patterns.
- The need to be housed with, or apart from, other animals.
- The need to be protected from pain, suffering, injury and disease.

Choosing a welfare assessment framework

The welfare assessment frameworks described above cover similar areas and are all useful for guiding thinking on animal welfare by veterinary professionals looking to improve the lives of animals (Table 1-2). They can all be used with the relevant caveats.

However, there are sometimes intense academic debates regarding slight differences in approach and semantics, which can filter into the real world and cause unnecessary confusion. For instance, the 'Five Domains' model has recently become popular with various institutions, and there is a tendency for it to be portrayed as very different from the Five Freedoms. Important arguments can be made for emphasising the similarities among the Five Domains, the Five Freedoms, and the other frameworks; in considering these, the Five Freedoms framework was deemed most appropriate for analysing welfare in the species-specific chapters in Part 2 of this book.

Since the publication of the Brambell report, a pivotal moment for animal welfare in the UK, animal welfare action has primarily been based on understanding and appreciating the broader needs of animals, as outlined in the Five Freedoms, which have gradually become internationally recognised. The Five

Table 1-2. The different welfare assessment frameworks comprise very similar concepts.

Five Freedoms	Five Domains	Welfare Quality®	Five Welfare Needs
Hunger and thirst	Nutrition	Good feeding	Suitable diet
Discomfort	Environment	Good housing	Suitable environment
Pain, injury and disease	Health	Good health	Protection from pain, suffering and disease
Normal behaviour	Behaviour	Appropriate behaviour	Exhibit normal behaviour patterns
Fear and distress	Mental state		House with, or apart from, other animals

Freedoms are useful in that they can be easily understood by people from many backgrounds, including those whose primary work is not animal welfare but who develop policy, legislation and corporate social responsibility strategies that impact animals. There are tremendous benefits to having a readily understood, globally accepted and consistent reference point for welfare, coordinating and directing action. These factors probably outweigh any perceived shortcomings in the Five Freedoms – after all, they can be used with relevant caveats in mind. Animal welfare developments take time, and continuity is essential for a unified message on animal welfare to remain heard and acted upon, avoiding sudden or confused changes in thinking that stall progress. Therefore, it is helpful that the leading international veterinary body, the World Organisation for Animal Health (WOAH), formerly the Office International des Epizooties (OIE), keeps the continuity by referencing both the Five Freedoms and the Five Domains, having used the Five Freedoms as their guiding principles in developing welfare standards from 2002 onwards.[9]

Relationship between welfare and 'naturalness'

Consumers often see a 'natural' life as important for welfare. The concept of naturalness does have value when assessing welfare. Animals have evolved to live in specific environments, and the further we remove them from their natural conditions, the greater the risk of poor welfare. This is because they become increasingly dependent on people to understand and provide what they need, rather than obtaining it themselves through choices they make within their natural habitats. However, to equate a 'natural' environment or 'natural' behaviour with good welfare is overly simplistic. For example, anti-predator, fear-related behaviours are natural but lead to poor welfare in captive environments, particularly if prolonged, and the animal cannot take actions that alleviate the fear. 'Naturalness' has been used to justify raw diets for dogs, which, despite certain benefits, can also cause harm due to poor cleanliness levels and nutritional balance of the food, as well as injuries due to bone fragments.

Issues related to naturalness are compelling for the public and consumers, such as cows grazing on grass, no cages and no mutilations. However, we need to look beyond this and ensure that the animal's viewpoint remains central, using evidence from animal welfare science (see Chapter 2). The human's idea of a desirable, natural environment, shaped by human preferences and values, must not usurp that of the animals. Instead, **the ability of the animal to adapt and cope with the environment in which it is placed**[10] is probably a more useful measure of welfare.

To illustrate, during a drought in Kenya, many farmers lost cattle. Donkeys survived as they were physiologically better equipped to cope with a lack of water. However, when donkeys were substituted for cattle (Figure 1-2), they were not adapted for the type of work they were expected to do, leading to wounds from the abrasive impact of the harness on their skin.

Figure 1-2. A donkey used for ploughing.

Part 2 of this book provides multiple examples of how the concept of 'ability to adapt' applies in different species and contexts.

A wider concept of animal welfare

Animal welfare is also concerned with effective action and outcomes for animals. David Bayvel's animal welfare definition[11] encompasses this 'bigger picture' for achieving real-world change. It recognises that animal welfare is an international and domestic public policy issue that must consider scientific, ethical and economic issues. There are also religious, cultural and international trade policy considerations – animal welfare issues must be considered alongside other important evolving issues, such as the growing emphasis on environmental concerns and sustainability. Multiple stakeholders are involved, including industry and producer groups, scientific bodies, animal-welfare non-governmental organisations and professional groups such as the veterinary and legal professions. David Bayvel and others recommend that change should be science-based, ethically principled and incremental. The rate of change should balance society's expectations and the constraints on the animal users, ideally with full 'buy-in' from the latter.

Animal Welfare Science

2

Abstract

This chapter explains how animal welfare has been studied scientifically and highlights key findings. The relationship between welfare and production is discussed. Special emphasis is placed on animal behaviour, as traditional veterinary training favours the physical aspects of welfare, such as health and disease. Examples include normal and abnormal behaviours, learning, human–animal interactions and human behaviour change. The scientific evaluation of sentience is considered.

Introduction

Key similarities between humans and other animals allow us to infer how they feel, particularly in species we can readily relate to and empathise with. For example, we infer pain in a mammal with injuries that would be painful in ourselves. Such anthropomorphism guided the initial development of animal welfare concepts. However, our use of anthropomorphism as a tool to understand animal welfare must be critically engaged with and evaluated. Many species experience the world differently from us and can potentially suffer or experience pleasure in ways we cannot due to their very different senses and perceptions. Therefore, the scientific study of animal welfare is vital for further clarifying what matters to different species in different contexts. In recent decades, animal welfare has become a well-established scientific discipline, with researchers using theories and techniques from different scientific disciplines, such as veterinary medicine, pathology, behaviour, physiology, and neuroscience, to answer questions about what animals feel and, increasingly, how to bring about change that improves welfare within relevant socio-economic limits. A vast body of scientific evidence now exists that can and should be used as part of the veterinary profession's evidence, and in developing policy and legislation concerning animals.

DOI: 10.1079/9781786391582.0002

Approaches to animal welfare science

Science can never be free of values (morals). Research has a cost; social, moral, political, and economic considerations ultimately determine what is funded. Ethics guides animal welfare research, including which species are deemed worth studying and which are ignored, what scientific methods are acceptable, and how animal welfare is conceptualised, including the relative importance that different scientists place on the mental, physical and 'naturalness' aspects of welfare and the intrinsic value they assign to different species.

In the UK, public concern over the welfare of farmed animals was one of the main impetuses for developing animal welfare science. The remit of the Farm Animal Welfare Council (FAWC) included using science when making recommendations about farm animal welfare. The range of species and contexts has gradually broadened from an initial focus on very confined farmed animals, such as intensively farmed pigs and chickens, to cattle, sheep, goats and fish. The welfare issues faced by common companion animal species have gradually been recognised. To date, studies on the welfare of wild animals are limited.

The field of welfare research has gradually moved from studying poor welfare to understanding how to provide animals with as many positive experiences as possible. Further developments include the 'allostasis' model of animal welfare, where animals are viewed as active participants in their environment rather than passive bystanders, recognising the importance of allowing them a degree of autonomy (choices).

When studying animal welfare, researchers draw inferences from three broad, interrelated categories of measures:

- Physical condition.
- Physiology.
- Behaviour.

They must be considered in conjunction with each other.

Each category is discussed in turn below. Additional emphasis is given to behavioural measures because veterinary professionals' training is often biased towards understanding the physical health of animals. Some general key findings are summarised, but they are by no means exhaustive: animal welfare is now a vast, multi-disciplinary field of research. Over time, there has been a move towards studying the positive aspects of welfare that help us decide how to provide animals with the best possible life, rather than simply studying negative aspects, such as pain and distress. Research options have also altered as new technologies become available and more affordable.

Measures of physical condition

Measures of physical condition include body condition, reproductive success, growth and mortality rates, and the type, severity, prevalence and incidence of injuries and disease. Veterinary professionals understand these parameters very well and automatically think of good health as essential for good welfare and poor health as causing poor welfare. However, the relationship between welfare and production is not straightforward.

Relationship between welfare and production

Welfare and production can improve in tandem. There are gains from improving husbandry and eradicating certain diseases to improve physical health. Unpleasant emotional states, such as fear of humans, can also reduce production, not just welfare, due to the activation of the body's stress response. High fear of humans has been shown to reduce egg production in laying hens, feed conversion efficiency and growth rates in meat chickens (broilers), and milk let down and milk yield in dairy cattle and goats. Fear of humans can also make animals more difficult to handle, increasing the risk of injury for both animals and humans.

However, this relationship between welfare and production has limits. The commonly heard statement 'my animals would not be productive if they didn't have good welfare' often does not hold. In numerous situations, animals can experience poor welfare but still be 'productive'. This is because production concerns the overall profits of the farming system and associated food chains. Production can often be increased by limiting input costs, such as reducing space allowance and staff numbers, which can lower animal welfare standards. Animals can be physically unfit but still produce well. Many modern-day production systems push animals to their metabolic limits. A whole area of veterinary medicine is devoted to the 'production diseases'. For example, broiler chickens have been genetically selected for a fast growth rate, selectively laying down breast muscle mass to the detriment of their general health. They can suffer from lameness and circulatory disorders, but still produce the commodity the marketplace demands, generally at a profit. Dairy cows can be lame but still produce the quantity and quality of milk the marketplace requires.

The horse in Figure 2-1 has poor welfare; he is thin and weak, with skin wounds and chronic musculoskeletal problems. However, a rubbish collector in Mexico City paid a low price for the animal and did enough work to make a profit, despite spending some of the earnings on feed.

Animals can be physically fit and productive, but still have poor welfare due to experiencing unpleasant feelings. For example, they can be bored, frustrated or fearful. Fast-growing pigs can be reared in barren environments that ignore their behavioural needs and still be productive. Social animals reared in isolation can appear physically healthy but have poor mental health and lack the social skills later needed to function well as part of a group. Therefore, **physical measures must be used in conjunction with other measures to gain a holistic welfare picture.**

Figure 2-1. Working horse in the rubbish dumps of Mexico.

Measures of physiology

Physiology concerns how a living organism or its bodily parts function. Animal welfare researchers commonly measure the biological changes that occur in the body in response to stress. Stressors can be either emotional or physical. Heart rate and blood pressure are commonly used to assess the circulatory response. White blood cells, hormones and cytokines in the blood can be used to evaluate the immune response. Blood glucose and non-esterified fatty acid levels can help measure the metabolic response. There are many more measures.

The primary body systems that react to stress – the hypothalamic-pituitary-adrenal (HPA) axis and the sympathetic nervous system (SNS) – have been well studied in various species. The HPA axis is essential for limiting inflammation and promoting the release of substances that repair damaged tissues. Activation of the HPA axis releases glucocorticoids, such as cortisol, from the adrenal gland into the bloodstream. The resulting cortisol levels in the blood, saliva, urine, faeces and milk can be measured. The SNS is essential for preparing the body for the effort required in defence or escape, that is, 'fight, flight or freeze'. When activated, heart rate and blood pressure increase, heart rate variability alters and plasma catecholamines (adrenaline and noradrenaline) are released into the blood.

However, physiological measures of stress are not straightforward indicators of welfare and must always be interpreted in context, along with physical and behavioural measures. The everyday usage of the word 'stress' implies that the animal is experiencing something unpleasant. In contrast, the scientific meaning of stress is that the body is simply aroused and preparing for action. Physiological measures alter when the animal is aroused by pleasurable experiences, such as the physical stress associated with play and the mental stress linked to excitement, not just unpleasant experiences. There is no precise measurement threshold at which we can say stress has escalated to become 'distress', that is, detrimental levels of stress that the animal has difficulty coping with. Factors unrelated to stress, such as age, sex, time of day and time of year, all influence their levels. In addition, techniques used to collect samples can cause stress and confound results, and researchers have had to devise ways to counter this. Chronic stress is even more challenging to measure than acute stress because relevant measures of HPA axis activity, such as glucocorticoid levels, rarely stay high for more than a few days due to efficient homeostatic feedback mechanisms. This is one of the challenges scientists have faced in developing measures of chronic stress.

Despite difficulties in their interpretation, physiological measures have proved very useful in animal welfare research. Examples include assessing and comparing the welfare impact of diverse husbandry practices, such as dehorning and disbudding, castration, tail docking, confinement, social dominance and social isolation.

Measures of behaviour

Understanding behaviours is essential to understanding welfare. Whilst animal behaviour research has a long history, it has only acquired an animal welfare focus in recent decades. Behaviour research is discussed below in three broad categories:

- Normal or natural behaviours.
- Abnormal behaviours.
- Learning and behaviour.

Normal behaviours

Normal animal behaviours can be broadly categorised as:

- Behaviours the animal wants to do much of the time, often internally motivated.
- Behaviours the animal wants to do some of the time, depending on environmental triggers.
- Behaviours the animal wants to avoid.

Animals have behavioural needs. They are highly motivated to perform some behaviours, regardless of the environment, and even if the end goal of the behaviour is already provided by the owner or carer. These behaviours are internally driven or instinctive and must be prioritised in the design of husbandry systems if welfare is to be protected. Examples include a pig foraging and exploring by 'rooting' or a hen nest building around the time she lays eggs. Scientific evidence shows that engaging in certain behaviours is rewarding in itself. For example, some species would rather 'work' to obtain food, performing specific foraging or problem-solving activities, even if the food is readily available to them without effort. This phenomenon is called 'contrafreeloading'.

Some behaviours are performed only if specific triggers are present in the environment. For example, pigs are only motivated to wallow to cool themselves if the ambient temperature exceeds a certain level (Figure 2-2).

Some animals lose the motivation to behave a certain way if the end goal of that behaviour is provided for them. For example, the motivation of gerbils to dig and create burrows may almost be abolished by providing them with a ready-made naturalistic burrow, provided it is of the correct design.

Some behaviours are 'luxuries' or optional, that is, they are only performed if other vital needs are met. These behaviours add to the animal's quality of life and are considered positive welfare indicators. An example is play behaviour (Figure 2-3).

Some behaviours must be performed at a specific time in the animal's life to prevent future welfare problems. An example is the socialisation of puppies during the 'critical period' (see Chapter 10).

Animals often want to avoid performing behaviours associated with aversive sensations and emotions, such as pain, fear, frustration and boredom. Examples include lameness to alleviate soreness in a limb, shivering to stay warm and excessive time lying inactive due to a boring environment.

Figure 2-2. Pig wallowing.

Figure 2-3. Goat kids playing.

Understanding normal behaviours

Studying animals in their natural environment

A helpful starting point for understanding normal behaviours is studying domestic animals free roaming in a naturalistic environment, and/or their wild counterparts, and documenting an 'ethogram' (catalogue of normal behaviour). An example is the pioneering work of Stolba and Wood-Gush, published in 1989,[12] who observed free-ranging domestic pigs in a semi-natural environment over three years, documenting behaviours and associated events and stimuli. Such studies have shown us that, despite thousands of years of domestication, many domestic animals retain the same behavioural traits as their wild counterparts. It is simply the threshold at which behaviours occur, and their frequency and intensity that have changed. For example, despite being domesticated, these pigs demonstrate extensive social and maternal behaviours such as building nests.

Ethological approaches can help us further understand these behaviours. Tinbergen's four questions provide a helpful framework by asking about 'causation' (What stimulates the behaviour?), 'development' (How does the behaviour develop during the animal's lifetime?), 'function' (How does the behaviour help the animal survive and reproduce?) and 'evolution' (How did the behaviour arise in the species, and why does it persist?). Animal welfare science asks the additional question: How does the animal feel if certain behaviours are allowed or denied?

Controlled studies linking animal feelings to their behaviours

Experimental approaches have been developed that essentially 'ask' animals what matters to them. Examples include preference and operant studies. In preference studies, animals are given choices, and their responses are measured. For example, they may be offered two or three types of flooring and the amount of time spent on each is measured and compared.

Operant studies use the principle that animals often want to access resources that will allow them to perform behaviours important to them. The strength of their motivation can be measured by equating it with the effort they are prepared to make to acquire that resource. They may be sufficiently motivated to perform an 'instrumental' response, such as pigs pushing levers to gain access to straw or hens pecking at keys to access food. The theory of consumer demand in humans has guided interpretation. Here, the more someone wants something, the more they are willing to pay for it. Certain products are regarded as 'necessities' and are purchased in approximately the same amounts irrespective of whether prices increase or incomes decrease. Others are considered 'luxuries' because their consumption falls as prices rise or incomes decline. Some behaviours in animals can be considered 'necessities'. For example, hens will work as hard for a nest site 40 minutes before laying as they will for food after being deprived for several hours, indicating that nest building before laying is as important to them as eating food when very hungry.

Behaviours and situations that animals want to avoid can also be identified by determining how hard they will work to avoid those experiences or what they are prepared to forego. For example, pigs are highly motivated to eat apples but will stop entering chambers containing a high concentration of gaseous carbon dioxide (CO_2) to receive an apple reward, showing they find this experience aversive. This finding has contributed to our understanding of welfare when slaughtering pigs using CO_2.

In all studies, instrumental tasks must be selected with care. The species and developmental stage must be naturally inclined to perform the task. Also, if animals become fearful, they may freeze or try to escape, leaving them unable to perform the required task.

Inducing specific feelings and observing the resultant behaviour

Animals are observed in situations known to induce specific feelings, and their behaviour is documented. Concurrent physical, physiological and behavioural measures, already validated, provide further evidence of the animal's mood. Animals are then studied in situations where the impact on their feelings is not known. If the same behaviours are shown as in the known situation, these are taken to indicate the animal is experiencing a similar feeling or emotion to that in the known situation. Existing animal welfare research has generally focused on fear, frustration and pain, with less study of boredom or other moods that may be equally important to welfare. Examples include exposing hungry animals to conditions where they have learnt to expect food but withholding it, creating frustration. Common behavioural responses are escape attempts, distress signals, aggression towards other group members, and irrelevant 'displacement activities'. If similar behaviour is subsequently observed in different situations and contexts, researchers can deduce that these situations are likely also to frustrate the animal. Other supporting evidence for frustration may include physiological changes, such as raised blood pressure and cortisol levels, as well as a reduction in behavioural signs after the animal receives anxiety-relieving drugs.

Similar approaches have been used to study fear and have shown how varied the associated behavioural responses can be. Humans can easily misinterpret fear behaviours. For instance, fearful animals may simply freeze rather than attempt to escape or bite. Welfare science has demonstrated that these animals are not 'relaxed' or 'hypnotised' as people sometimes presume. The chicken or rabbit that remains immobile for a time period after being held on its back, a phenomenon known as 'tonic immobility', is fearful. This behaviour might be 'feigning death' as a strategy to escape predators.

'Listening in' on animal communication to identify pain and distress

Animal welfare researchers have developed techniques to 'listen in' on the communication of some species. For example, studies have recorded piglet vocalisations in different contexts, such as when they are removed from their mothers or are having their tails docked. Knowledge of the context is used to interpret what their different calls mean. Similar calls are then listened for in other situations where the welfare impact on the piglet is less known.

Environmental enrichment

Effective environmental enrichment improves the lives of captive animals, making them more interesting and rewarding, by offering greater choice, a wider range of behavioural expression, the use of their problem-solving skills, and opportunities to engage in positive social interactions. Alterations can be physical structures, scents and smells, methods of food provision and carefully chosen social groupings. A well-chosen combination is likely to maximise the welfare impact. Effective strategies will be species-specific, provided in sufficient quantities for all animals, and practical, affordable and safe. They will maintain the animals' interest and provide the correct level of challenge – too much novelty and change can induce anxiety and fear. Species-specific examples of enrichment are provided in Part 2.

Abnormal behaviours

Abnormal behaviours are behaviours not found in the animal's normal ethogram or natural behaviours that are performed excessively. They generally indicate poor welfare, with the animal's environment deficient in something it needs to function normally. Note that some unnatural behaviours can be considered normal in some contexts (see 'Learning and behaviour' p. 20).

In animal welfare science, definitions for abnormal behaviours seem to have arisen piecemeal, with different researchers using different approaches. Examples include redirected behaviours, vacuum activities, displacement activities and stereotypic behaviours.

Redirected behaviours

Redirected behaviours occur where the environment lacks something the animal would normally use to express a highly motivated behaviour, so they direct the behaviour at something inappropriate. For example, pigs will bite the tails of other pigs when kept in barren environments as they lack straw and other manipulable materials to which they would normally direct their highly motivated exploratory and rooting behaviours.

Vacuum behaviours

Vacuum behaviours are analogous to an animal miming a behaviour where it lacks the resources to perform it properly. An example is vacuum chewing in pigs, where they perform all the activities associated with chewing but have no substrate in their mouth.

Displacement activities

Displacement activities occur when animals are in conflict or frustrated, through being motivated to act in two different ways simultaneously or being prevented from performing something they expect to be able to do. An example is hungry hens vigorously preening themselves when prevented from accessing their routine food source.

Stereotypic behaviours

Stereotypic behaviours have been particularly well studied. These are repetitive, invariant behaviours that have no obvious goal or function. Often, the behaviour is predictable, occurring at the same location or time of day. The animal may have difficulty stopping the behaviour, even if the root cause is remedied. In contrast, 'habits' are generally learned, behavioural routines, often triggered by specific cues, that can be functional or adaptive. There are many different types of stereotypies. Oral stereotypies may develop from redirected forms of sucking, eating or foraging behaviour. An example is chain chewing in sows. Locomotor stereotypies may develop from frustrated locomotion. An example is when bears repeatedly pace around their cages in zoos. It is unclear what, if anything, the animal gains from performing stereotypies. Certainly, they are a sign of poor welfare, and they can further reduce welfare by limiting the time animals spend on normal behaviours that are important for their well-being.

There may be underlying brain pathology. For instance, administering amphetamines to experimental animals has been shown to induce stereotypies, suggesting they are due to similar mental disturbances experienced by humans when taking amphetamines.

However, stereotypic behaviour could also be part of a coping strategy, a type of 'do it yourself' enrichment. Where individuals are kept together in a poor welfare environment, those performing stereotypies often have better welfare than those who do not. For example, persistent tongue rolling in veal calves can stimulate saliva production, buffer stomach acidity and lead to fewer stomach ulcers (ulcers are associated with a diet lacking forage), compared with calves that do not perform this abnormal behaviour.

Simply taking steps to prevent the stereotypical behaviour without addressing the underlying cause can worsen welfare. Examples are preventing the abnormal behaviours of crib-biting and wind sucking in a horse by using a cribbing collar or surgically resecting nerves and muscles to restrict the horse's neck stretching ability, rather than correcting underlying dietary and other factors.

Mutilations

Mutilations are procedures that interfere with the sensitive tissues or bone structures of an animal for purposes other than its medical treatment. Examples include disbudding calves and goat kids, desexing of farm and companion animals, tail docking in pigs, lambs and dogs and beak trimming in poultry. Many are further explained in Part 2. Mutilations broadly involve surgically altering animals to suit our environmental and husbandry choices, rather than adapting the environment and management to accommodate the animals' normal behaviours. In some species and contexts, mutilations are routinely performed to limit the production harms due to abnormal behaviours rather than addressing the root cause of the deficiency in the animal's environment. In farm animals, many procedures are performed with no or limited analgesia and anaesthesia. A full ethical discussion is outside the scope of this book, except to say that decisions should be made on a case-by-case basis, considering not only how to improve animal welfare during the procedure but also questioning the necessity of the procedure itself.

Learning and behaviour

The ability of animals to learn is crucial for their welfare. Many behaviours we want and expect from the animals we keep are unnatural to them. For instance, it is not 'natural' behaviour for a prey species such as horses to stand still and lift their feet for shoeing by a farrier or wear tack and be ridden. Likewise, it is not natural for cattle to enter a milking parlour and allow a human close contact and to attach milking machinery to their udders. The more instinctive and natural responses of untrained animals would be to resist or escape. Desired behaviours must all be learned. It is easy to take this for granted because so many do cope. The ones that do not cope are often unfairly labelled as having 'behaviour problems'.

Learning involves acquiring knowledge or skills through practice and experience. Animals constantly learn from their surroundings, including their interactions with other animals and humans. They learn whether we intend to teach them or not – it is their experiences that matter, not our intent. Animal learning has been extensively researched in various scientific disciplines, though interest in its welfare implications is relatively recent. Many animals learn in similar general ways to humans. However, different species have unique perceptions and natural tendencies that affect their learning. People will often not detect many of the cues animals pick up on through their very different senses of smell, hearing and vision. However, we can be mindful of these species-specific differences and apply this knowledge when interacting with and training animals, thereby creating the best possible experience for them, which often results in a rewarding experience for us as well.

Categories of learning

Learning has been broadly divided into non-associative and associative categories.[13] There has been much less scientific evaluation of complex learning, such as thinking, planning, social and observational learning. This does not mean these types of learning do not occur; it is just that they are more challenging to study scientifically.

Non-associative learning

Non-associative learning is when an animal's response to a single stimulus (cue) changes over time with repeated exposure. Habituation is when the response decreases until it disappears altogether. This mechanism helps individuals filter out insignificant events and focus on those that are potentially dangerous. For example, a horse may initially be startled by the sound of a passing train when grazing in a new field next to a rail track, but will eventually habituate to it with repeated exposure. Sensitisation is the opposite of habituation, where the response intensifies with repeated exposure, for example, the fear some dogs have of loud noises such as thunder or fireworks. Desensitisation is a technique that behaviourists use in a controlled setting to help animals overcome their fears, for example, those in dogs associated with loud noises. A stimulus that usually causes alarm is introduced at a very low intensity that does not trigger a fear behavioural response. The intensity is gradually increased as the animal adjusts, always keeping it below the threshold that could trigger a fear reaction.

Associative learning

Associative learning is when an animal associates two seemingly unrelated stimuli by experiencing them together. The two main types of associative learning are classical (Pavlovian) conditioning and operant conditioning.

In classical conditioning, an animal learns to associate two stimuli that are not naturally paired. Pavlov's experiment demonstrated that by repeatedly ringing a bell when presenting food to a dog, the dog eventually learned to associate the bell sound with food and would drool at the sound alone. Similarly, sheep will follow the sound of a feed bag shaken by the farmer as they have learnt to associate this with feed. Consider a horse receiving an intravenous injection. Whilst a hand used to raise the jugular vein is initially a neutral stimulus, it only takes a few repetitions before the horse predicts venepuncture. Raising the vein then evokes the negative emotional response, previously only evoked by the needle. This becomes problematic, as when raising the vein evokes a fear response, the horse is now primed to react aversely to venepuncture, i.e. the animal has learnt to predict what will happen next and to react based on this prediction.

Operant conditioning is a process where an animal learns to associate a particular action with a specific outcome. The animal performs a certain behaviour and experiences the consequence. The nature of the consequence – whether liked or disliked – determines how likely the animal is to repeat that behaviour in future (Table 2-1).

Veterinary professionals often actively use these principles and can educate others on how to use them to optimise the welfare of the animals they interact with. Sometimes, people unintentionally reinforce or punish certain behaviours, creating problems.

The timing of the reinforcement is critical, as the association between the event and outcome is often made in under half a second. Too long an interval, and the animal does not make any connection. A range of examples is given below. However, note that training should be based on positive reinforcement wherever possible, which involves providing the animals with a reward. Positive punishment should not be used.

Examples of positive reinforcement

- Animals learn that if they stand still for us, they will immediately be given a food reward.
- An animal fiddles with a gate latch and finds that the gate opens. The animal learns to repeat this behaviour due to the enjoyment and interest created by escaping and exploring.[13]

Table 2-1. Reinforcement and punishment: 'Reinforcement' makes the behaviour more likely to occur. 'Punishment' makes the behaviour less likely to occur. 'Positive' means something is added. 'Negative' means something is taken away.

	Likes/wants	Dislikes/avoids
Add	Positive reinforcement	Positive punishment
Remove	Negative punishment	Negative reinforcement

Examples of negative reinforcement

- Riders increase their leg pressure to make horses go faster. This causes mild discomfort, but the pressure is released as a reward when the horse moves. The horse learns that moving forward avoids discomfort and increases pace.[13]
- If a person enters a cow's flight zone, then the cow experiences discomfort. By moving away from the person, the cow lessens her discomfort.
- Animals may learn to escape a procedure they dislike once they realise that the escape behaviour immediately lessens their discomfort.

Examples of positive punishment

- A person slaps an animal to try to get them to stop the behaviour they are engaging in. Positive punishment, especially when combined with poor timing, often leads to poor welfare. The animal is likely to become fearful and fail to make the connection between cause and effect that we desire. A vicious cycle of repetition can follow as the owner becomes increasingly frustrated.
- A horse receives a painful injection, perceives it as punishment, and due to the association, becomes 'needle shy'. Animals do not necessarily need much repetition to make associations.[13] From then on, the horse tries to avoid needles or syringes by using escape behaviours, many of which will be successful due to the horse's large size and power.

Examples of negative punishment

- A horse flattens her ears back when a person starts to enter the stable with a bucket of feed. To discourage this behaviour, the person stops the process and walks away with the feed before the horse receives any.[13]

Shaping

Shaping is the process used to teach more complex behaviours that the animal is unlikely to perform spontaneously. It involves breaking down the end behaviour into several easily achieved steps. The trainer needs to gradually move the animal closer to the desired behaviour by rewarding behaviours that gradually approximate it, until the desired behaviour is eventually performed in full. Consider the 'needle-shy' horse. Letting go of the horse when he rears inadvertently rewards that behaviour. However, waiting for the horse to stop rearing and remain still is difficult, dangerous and inefficient. Instead, you can gradually shape the desired behaviour. For instance, start by placing your hand on the horse's shoulder. If the horse stands still, reward that behaviour by removing your hand. To enhance the reward, provide a small treat simultaneously. Repeat the above, each time moving your hand closer to the horse's vein, continuing to reward the behaviour of standing still. The speed of progress will depend on the horse's levels of arousal or alertness. If you place your hand on the horse's shoulder and he shifts his weight or raises his head, you are probably approaching the limit of what he is comfortable with. In this case, repeat the initial step several times until the horse appears more relaxed before attempting to move your hand closer again.

Clicker training

Classical and operant conditioning are often used together. An example is clicker training. The click sound starts as a neutral stimulus. The animal learns to associate the sound of the click with a food reward, which is a 'positive reinforcement', through classical conditioning. Soon, the animal learns that the click sound predicts the arrival of food, and with repetition, the click sound itself starts to create the same positive emotional response as receiving food. The interval between the click sound and the food reward is gradually increased so that the click sound becomes a 'bridging stimulus' between the desired behaviour and the food reward. The click sound is useful as it can be more accurately timed and, therefore, a better reinforcer for the desired behaviour than using a food reward only.

Generalising

Animals can learn to generalise their behaviour and respond in the same way to similar stimuli in different locations. For instance, pigs can learn to associate a particular colour of overalls with gentle handling and respond more positively to anyone wearing the same colour overalls, even in a different place. However, generalisation is not automatic. For example, a horse that has learned to jump over an obstacle at home may struggle to do so in the different setting of a showground.

Human behaviour

There is a whole area of research devoted to human–animal interactions and human behaviour change, as the people who care for animals – whether called 'stockpersons' for farmed animals, 'keepers' for zoo animals, or 'owners'/'guardians' for companion animals – are often the most important determinant of their welfare.

Human–animal interactions have been most extensively studied for farm animals, although findings are relevant to all animal carers. Important stockperson traits have been broadly categorised as 'willingness', 'capacity', and 'opportunity'.[14] Willingness covers the desire to look after animals well, motivated by empathy, a genuine concern for animals and pride in the job. Research topics include the stockpersons' attitudes toward animals and their underlying beliefs, how these shape interactions, and how attitudes and behaviours can be modified through training. Capacity covers the capability of stockpersons to perform their job, including knowledge levels, practical skills, experience, and physical and mental fitness. Understanding animal behaviour and welfare, and respecting this in interactions, particularly when handling animals, is crucial. Many additional skills, such as nutrition, biosecurity, and working with machinery, are also required. 'Competency' is an everyday term that encompasses both willingness and capacity. People also need the opportunity to care for the animals well, through working in an environment and management system that provides sufficient time and resources, designed to promote both animal and human well-being.

Taken together, willingness, capacity and opportunity determine the behaviours of the animals' carer. Where all three are optimal, people can provide care that meets the 'welfare potential' of a husbandry system. Shortcomings in any area will lower welfare standards. Even a highly motivated, skilled, physically fit carer will struggle if given too little time or resources. Conversely, the very best facilities cannot compensate for poor motivation and skill sets.

Some husbandry systems have 'inherent' welfare constraints. These are the welfare issues that remain even when the husbandry system is managed optimally. For example, it may be the industry norm to have very large numbers of animals and few people caring for them, making it very difficult to properly monitor individual animals and interact with them in ways that do not frighten them. It may be the industry norm to house animals in ways that severely restrict their behaviours or use genetic strains at high risk of pathologies. Competent animal carers can mitigate but not neutralise these harms. This is why 'method of production' labelling can be useful for indicating the welfare provenance of food and other animal products, provided it is accompanied by direct assessments of the animals using 'animal-based measures', verifying that the level of care is acceptable (see Chapter 3 for further explanation).

Human behaviour change (HBC)

Veterinary professionals often have to ask others to change. For instance, we often have to ask people to provide additional care for their animals that they have not given before or to change management practices they have used for years. While most animal owners and carers want the best for their animals, this does not always translate into action. It is not enough to follow the traditional route of providing knowledge alone and hoping that change will occur.

Human behaviour change research provides a deeper understanding of human behaviour. It draws on multiple disciplines, from the social sciences through to technological innovations. Below are examples of study findings.

Learning and memory

Humans, like other animals, learn by forming associations. For example, when people achieve a task and receive recognition for it, this can be rewarding and reinforces the behaviour involved. They often require support to help them succeed. Research suggests that using participatory education and training, where a skilled person – such as a member of a veterinary team or an experienced caregiver – demonstrates the desired behaviour and then provides ongoing support, can be highly effective. Repetition helps to practise the behaviour, making it more routine. Experiencing success can help individuals develop a 'growth mindset', where they remain open to learning new things throughout their lives.

Self-efficacy

Belief in one's ability to succeed in a specific situation or accomplish a task – known as self-efficacy – is crucial when making changes. What may seem simple to one person can feel complex and overwhelming to another. Supporting self-efficacy, alongside other key factors, can greatly enhance the likelihood of success.

The power of habit

Many aspects of animal care are routine and involve habitual behaviours. Developing positive habits, such as conducting regular health checks, while moving away from negative habits, like hitting animals during handling, has significant potential to create sustainable change. Habits tend to be more robust than intentions, because habitual behaviours require little conscious mental energy and motivation. Intention alone is often insufficient to bring about change.

Communication

Effective communication is crucial. The message should be clear, supportive and motivating. It is essential to employ an empathetic communication style that incorporates active listening and reflection, ensuring the animal carer feels heard and the situation is clearly understood.

Assessing what the animal carer does and does not understand is vital. For example, someone may feel disenfranchised and misunderstood, which can create barriers to change.

Expressing shared values, such as the goal of doing what is best for the animal, is generally more effective and motivating than simply providing the owner with a list of instructions. Supporting someone in coming up with their own solutions is often more powerful; people typically do not respond well to being told what to do and are more likely to continue a behaviour if they feel involved or believe they generated the idea themselves. Positive framing is helpful, focusing on empowering individuals by highlighting opportunities for change rather than dwelling on negatives.

Emotion versus rational thought

People often believe they are rational beings who carefully consider their decisions, but in reality, they frequently rely more on emotions than on rational thought when making choices. For example, a pet owner may give their pets too many treats simply because it makes them feel good. Emotional decision making tends to be the default unless we engage in deliberate thought, as it is easier and requires less cognitive effort. After all, this is what we evolved to do on a daily basis to help us survive.

Social influence

Providing social support is particularly powerful when changing behaviour. Understanding current social influences is important. These can be beneficial or detrimental. A social group that practises a desired behaviour can be a particularly strong support for positive change, while a group that does not can create obstacles. For instance, if someone belongs to a group that employs outdated dog training methods, they may require substantial support to challenge the group's established beliefs and practices.

Motivation

Behaviours driven by intrinsic motivation are more sustainable than those driven by extrinsic motivation. Intrinsically motivated actions are performed out of enjoyment or emotional satisfaction; for example, an owner feels good when providing care for their animal. Conversely, extrinsically motivated behaviours are performed for external rewards, such as receiving payment, and tend to have limited duration and strength.

Willpower

Activities generally become easier with repetition. However, willpower and energy are needed to perform that initial repetition. Fatigue can set in, meaning the person does not do the behaviour. They then often feel like a failure, making it is difficult for them to restart.

HBC research offers a range of approaches, ideas and expertise for designing effective HBC interventions. Such interventions support people in changing their behaviour in various contexts, whether as individuals, groups, communities or society as a whole. Intervention selection depends on the specific goals and behaviours required, as well as understanding what drives current behaviours and what is needed to facilitate change in different individuals. While veterinary professionals already implement some of this HBC knowledge, there is often room for further skill development. A conscious effort is needed to stay focused, particularly given the considerable time pressures faced during veterinary consultations. Developing successful HBC strategies may require external expertise and take time, but the investment is worthwhile for achieving sustainable improvements in animal welfare. The well-being of the caregiver/owner often improves alongside. It is encouraging to see that communication skills are increasingly being taught in a structured, evidence-based manner in veterinary medicine. Ideally, other aspects of HBC will also become more mainstream. Many of the advocacy examples in Part 2 could probably be made more effective by incorporating insights from the HBC discipline.

Resources providing further insights into what the HBC discipline can offer include Human Behaviour Change for Life (HBCL) https://www.hbcforlife.org

Sentience: further considerations

Broadly, sentience concerns the capacity to experience sensations, feelings or emotions, both pleasant and unpleasant. It relates to consciousness, something many of us intuitively understand. However, consciousness has proved tricky to define scientifically, other than to say it includes the following elements: perception of the environment through various sensory receptors, such as those that respond to light, sound, temperature, pressure, touch and so on; being able to process these sensations; the ability to experience pleasant and unpleasant emotional states and self-awareness.[15] Sentience and intelligence should not be conflated. Animals are more likely to be considered sentient if they are intelligent, that is, able to learn much, learn fast and make few errors once they have learned. However, good cognitive and learning abilities are not prerequisites for sentience; they will affect how an animal can suffer, but not whether it can suffer. In some situations, animals that are less able to learn and alter their behaviour based on learning may suffer more, as they repeatedly return to painful or unpleasant situations.

We generally agree that our veterinary patients and many other animals are sentient. However, we are still striving to clarify the developmental stage at which they become sentient. We are also trying to clarify where sentience arises on the phylogenetic tree.[16]

Science cannot provide absolute proof that animals, or indeed other humans, are sentient. It can only make inferences using the best available evidence. Much of this to date concerns the perception of adverse stimuli, particularly those that cause pain. Important research questions include:

1. Does the animal or species being studied possess the physical structures and physiological systems necessary to experience pain and other sensations and emotions?

2. Does the animal or species being studied show a direct behavioural response to pain, such as withdrawal from the painful sources or altered behaviour in the face of a painful stimulus it cannot escape?
3. Does administering appropriate analgesia modify this behavioural response to pain?
4. Can the animal learn from such experiences and minimise and avoid painful events in the future?

There are reasonable grounds to believe that all vertebrates (mammals, birds, fish, reptiles and amphibians) are sentient. In addition, there is significant evidence to believe that certain invertebrates are sentient – cephalopods (octopus, squid, cuttlefish, nautiloids), decapod crustaceans (prawns, crabs and lobsters) and cyclostomes (lamprey and hagfish).[17]

Many vertebrate animals have a pain system quite similar to ours, involving receptors sensitive to noxious stimuli, neural pathways, physiological responses and analytical brain centres. For instance, fish have nociceptors and neurotransmitter substance P (a neurotransmitter typical of mammalian pain systems) and produce opioids for endogenous analgesia. They show behavioural responses to pain, such as resting on the substratum, rocking from side to side, and rubbing their mouth on solid objects, when irritant solutions are applied to their mouth. This behaviour stops when analgesics are given. Many invertebrates also share these features. For example, cephalopods, such as octopuses, have nociceptors, the sensitivity of which increases after tissue injury, which helps them to avoid further damage. They have a nervous system, pain neurotransmitters and relatively complex brains. They release adrenal hormones in response to situations that elicit pain and distress in humans. Their colour changes rapidly in situations known to elicit pleasant and unpleasant feelings in many sentient animals, and they respond to the colour changes of others in the group. They have good learning ability and memory retention, readily learning to avoid stimuli generally considered painful in humans, and even learning to avoid stimuli that are not in themselves unpleasant but have been previously associated with an aversive electric shock.

Whilst the cerebral cortex is the seat of consciousness and learning in humans and many mammals, other taxonomic groups use different brain areas for these functions. The striatum part of the brain controls the exceptional learning and cognitive abilities of birds. Fish are a vast taxonomic group with very diverse brain structures. Some cephalopods and tetrapods do not have a central brain, instead possessing multiple peripheral ganglia. UK law recently recognised cephalopods and decapod crustaceans as sentient (s5(1) Animal Welfare (Sentience) Act).

There is currently less evidence of sentience in invertebrates such as insects. For example, salticid spiders possess sensory processing abilities and exhibit good analytical and predictive cognitive skills, but there is little evidence of a pain system. Further developmental research may help to determine the level of sentience.

Decisions on how we treat animals should be based on an understanding of their welfare needs. Therefore, the precautionary principle is promoted for many mammals, birds and increasingly in fish. However, there is still so much we do not know. It appears impractical to apply the same precautionary principle for many of the species we currently consider to have low sentience levels, whilst awaiting further scientific evidence.

Scientists are still trying to establish the stage of the animals' development at which sentience arises. Current evidence suggests that many vertebrates are capable of sentience during the last third of their development period, when still inside the mother or the egg. Sentience may develop earlier in precocial than in altricial species. Precocial species' offspring are much more mature and mobile immediately after birth or hatching, so they must be in a relatively advanced stage of development by the time of these events. Very precocial species tend to be oviparous rather than viviparous. For instance, some reptiles may develop conscious awareness hours or days before hatching. For some birds, such as the domestic chick, sensory and neural development is well advanced several days before hatching, with some responding to touch and light and even communicating with each other through clicking sounds known as 'pipping'. Some even breathe for several hours before they hatch. Therefore, protecting the welfare of many developing birds, reptiles, amphibians, fish and cephalopods might be more urgent than protecting many prepartum mammals. Further research is needed.

Whether developing animals become sentient before parturition or hatching may depend on the activation of various pathways that support consciousness. For species that breathe air, there is evidence that breathing is an important trigger for consciousness. Prepartum lambs have been extensively studied, as sheep have been the preferred laboratory animal for research examining risks during parturition in humans. Studies show that several endogenous factors suppress the consciousness of lambs *in utero* until oxygenation levels rise.[18] Whilst transient or episodic periods of conscious awareness *in utero* cannot be ruled out, evidence from electroencephalograms suggests they do not occur. Instead, the foetus alternates between two sleep states, rapid eye movement (REM) and non-REM. Such knowledge has practical implications for veterinary professionals, guiding the handling of the gravid uterus during surgery or when pregnant animals are euthanised. Current advice is not to open the uterus of a pregnant mammal in the last third of gestation until at least 30 minutes after their death, to be sure foetuses have died.[19] Therefore, they will not risk breathing air and becoming conscious.

Ideas on the sentience of neonates have changed in recent decades. Up until the 1970s, it was thought that neonates, including newborn babies, did not have a fully functional pain system and so could not experience pain, which led to operations on human infants being performed without anaesthesia and analgesia. However, it is now known that the incomplete myelination of the nervous system in neonatal animals does not prevent them from feeling pain but instead prevents the transmission of descending signals that inhibit pain, meaning they are likely to feel as much, or even more, pain than adults.[20] Pain behaviours are simply different in neonates, being more generalised than in adults. Also, painful experiences in neonates can lead to a general, long-term pain hypersensitivity. The earlier misconceptions have guided animal welfare legislation that allows us to perform mutilations such as castration, tail docking and disbudding on neonatal farm animals without using anaesthesia and analgesia. Operations at this age are often quicker and more straightforward to perform than in adult animals, with less haemorrhage and easier restraint, but they are unlikely to be less painful than the equivalent procedure on an adult animal.

Practical Welfare Assessment 3

Abstract
This chapter explains how animal welfare concepts and science have been combined to create credible, practical welfare assessment protocols for different species and contexts. It highlights existing protocols that veterinary professionals can use to improve their welfare assessments. This information should place them in a stronger position to contribute to further developments, refining existing protocols or creating new ones. The role of protocols in raising animal welfare standards is illustrated using examples from farmed animals. Scoring systems created specifically for assessing pain are then discussed.

Introduction

It is essential for veterinary professionals to have strong practical skills in welfare assessment. These skills are crucial for monitoring the effectiveness of health and welfare plans, providing input on the creation and implementation of welfare assurance programmes, assessing the effects of changes in animal husbandry systems, and evaluating compliance with animal welfare laws. Using credible welfare assessment protocols can improve assessments and make the process transparent to all stakeholders.

Features of ideal welfare assessment protocols

For welfare assessment protocols to be meaningful, widely trusted, accepted and used, they need to encompass several key features:

- They will be shaped by credible welfare assessment frameworks, such as the Five Freedoms, Five Welfare Needs, Five Domains or Welfare Quality® (see Chapter 1), and they will be underpinned by science.

DOI: 10.1079/9781786391582.0003

- They will cover the full life cycle of the animal, as welfare potential can vary greatly at different stages within the same farm. For instance, on pig farms, adult sows may be housed in straw yards, but the piglets and growers are reared in barren environments.
- Expert opinions, shaped by both scientific knowledge and practical experience, are crucial in selecting measures.
- They are best developed in conjunction with stakeholders, such as farmers, owners, veterinary professionals and consumers, because they all provide valuable insights. For instance, identifying 'win-win' situations on farms, where welfare and production can improve together, is particularly helpful for achieving welfare improvements based on feedback from assessment results.
- Outcomes should be measured, not just inputs. Inputs are the resources provided to animals that determine the likelihood of their having good welfare. Examples are feed, water, stocking density and the skills of people caring for the animals (see Figure 3-1). Many inputs, such as space allowances and stocking density, can be readily observed and measured. Others, such as stockmanship quality, can be harder to assess. Inputs alone do not provide a complete welfare picture. For this, outcomes, also called 'animal-based measures', are needed. These are direct observations of the animal's condition and comprise both physical and behavioural signs.
- Measures should be valid, repeatable and feasible. Valid measures are those that accurately measure the aspect of welfare they purport to measure. Repeatable (reliable) measures give consistent results, both when a single assessor measures the same thing on different occasions (intra-observer repeatability) and when different assessors measure the same thing (inter-observer repeatability). Feasible measures are practical to perform with the time and resources available.

Many measures used to assess welfare in controlled experiments, such as blood levels of inflammatory mediators and cortisol, are not practical for real-world use. However, evidence from these studies has validated many of the measures included in practical welfare assessment protocols.

It is helpful for policy reasons if the results allow welfare standards to be graded from poor to good, allowing different establishments to be easily compared. However, this is trickier than it sounds because it involves value judgements about acceptable levels of welfare, and people vary in the importance they give to different aspects. For instance, the Animal Welfare Committee has extensively discussed classifying animals as having a 'good life, a life worth living or a life not worth living'.[21] Despite these difficulties, results are still useful for comparing establishments with similar systems and analysing why some are meeting their welfare potential but others are not.

Stakeholders must be able to trust assessments. Therefore, independent and robust audits by skilled, trained assessors are important, ensuring fair comparisons of different establishments, particularly where welfare assessment is a mandatory part of an assurance scheme.

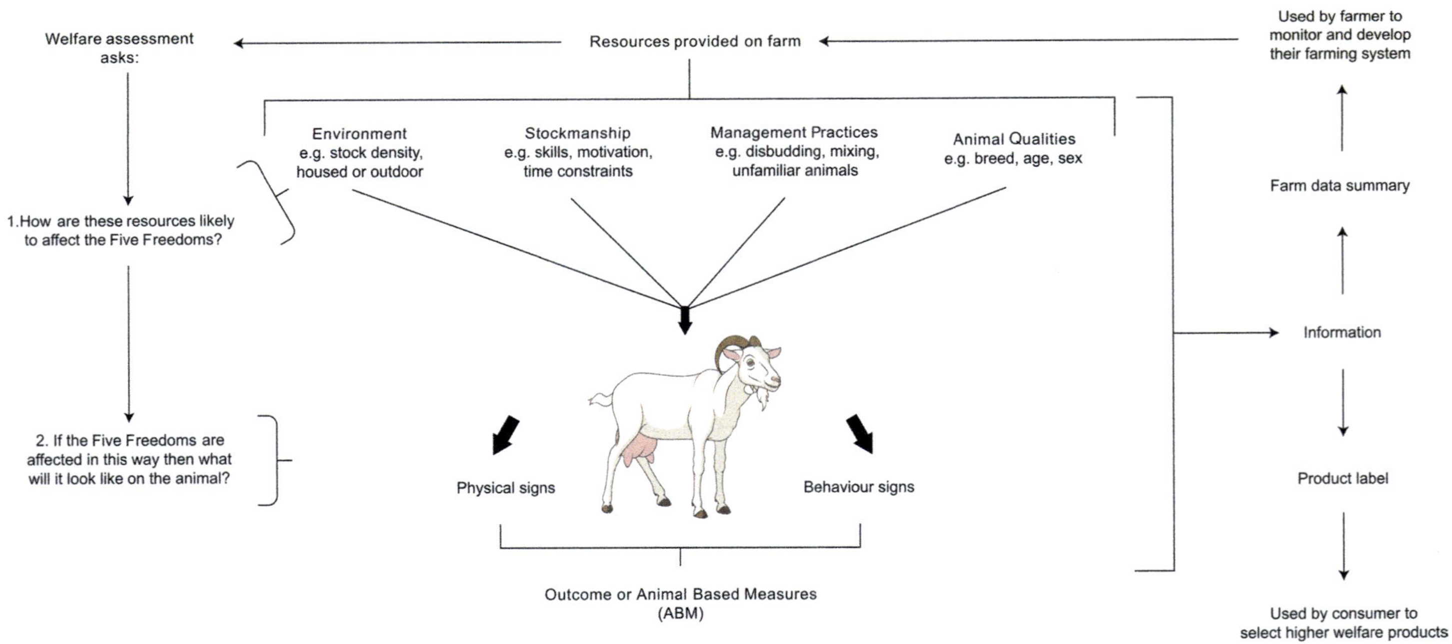

Figure 3-1. The main inputs (resources) and outcomes used in animal welfare assessment and their relationship with the 'Five Freedoms'.

Protocols for farm animals

Projects such as Welfare Quality®,[8] the Animal Welfare Indicators Network (AWIN),[22] and AssureWel[23] have developed protocols for commonly farmed species in Europe, incorporating the key features set out above.

Welfare Quality®

Between 2006 and 2010, the European Union funded various research projects, collectively under the umbrella of Welfare Quality®, that created preliminary practical welfare assessment protocols for farmed cattle, pigs and poultry. Protocols are based on the four welfare principles and 12 welfare criteria of the Welfare Quality® framework (see Chapter 1). Measures are mostly outcomes (animal-based). Based on results, farms can be placed in one of four categories, from poor to good. One of the project's aims was to actively improve welfare by communicating welfare standards to farmers and consumers, helping both to take action to support better welfare.

Animal Welfare Indicators Network (AWIN)

Welfare Quality® was followed by the Animal Welfare Indicators Network (AWIN) project, which developed preliminary protocols for the less commonly farmed species of sheep, goats, horses, donkeys and turkeys using a similar approach. This project was also funded by the European Union.

AssureWel

Protocol development is an iterative process that incorporates feedback from real-world use and evolving scientific evidence. Therefore, many protocols are still a work in progress. AssureWel is a good example of further refinements. AssureWel began as a UK-based collaborative research project between the University of Bristol, the Soil Association and the RSPCA between 2010 and 2016, aiming to facilitate the use of animal-based measures when assessing welfare and, therefore, improve the welfare assessment aspect of several well-established general farm assurance schemes.

It used learning from the Welfare Quality® project to create more straightforward, user-friendly protocols that remain fit for purpose and are adapted specifically for UK farming systems. As a result, AssureWel protocols are now embedded to varying degrees in several farm assurance schemes such as Soil Association Certification, RSPCA Assured and Red Tractor.

The AssureWel approach has four key elements, all needed to improve farm animal welfare:

1. Development of welfare outcome assessment protocols that are fit for purpose.
2. Training and standardisation of assessors to ensure competence and consistency.
3. Use of assessment results to refine farm management and improve animal welfare.
4. Engaging with industry to maximise impact.

Protocols prioritised measures for areas of high welfare concern, measures that aligned with each assurance scheme's animal welfare goals and measures sufficiently valid and reliable for purpose. AssureWel assessment protocols cover individual animals and groups of animals, as well as farm records (Table 3-1). They are available from www.assurewel.org, along with further explanations of measures and forms for recording. Assessors must receive appropriate training in using the developed protocols so they are competent and confident in implementing them on farm, and applying them consistently. AssureWel runs

Table 3-1. Summary of AssureWel measures for laying hens and pigs.

Category	Laying Hens	Pigs (Finishers/Dry Sows)
	Group	
Selection	All birds in a flock	Pigs in 4–5 randomly selected pens
Assessment measures	Beak trimming Antagonistic behaviour Flightiness Birds needing further care	Enrichment use Tail docking Ear/flank biting Pigs needing further care Lameness Hospital pen
Record-based measures	Mortality	Mortality
	Individual	
Selection	50 birds from the oldest flock	Up to 50 sows or finishers across a herd
Assessment measures	Feather loss (head/neck, back/vent) Dirtiness	Body marks Tail lesions Shoulder lesions Vulva lesions Manure on body Leg swellings Skin condition Lameness (finishers) Body condition

training programmes for all available protocols, which include classroom, farm and online testing elements.

Several working practices facilitate assessors in reviewing the results and identifying which management practices should change, and communicating with farmers. For instance, simple feedback forms are provided to promote positive discussions between assessors and farmers. Farms can be readily compared with other farms through benchmarking of results, and there are threshold levels of measures below which action to improve welfare should be triggered. Farmers are supported by ensuring they have viable solutions through knowing where interventions have been successful on other farms and where advisory support and technical guides can be accessed. The AssureWel project recognises that involving and engaging farmers and the wider food and farming industry is essential when developing, implementing and acting on the results of welfare assessments. For example, it engaged with other industry audit processes already in place by using the industry-adopted Agriculture and Horticulture Development Board (AHDB) mobility scoring system for dairy cows (Figure 3-2) and worked with AHDB to produce other resources.

Paying for higher animal welfare standards

Rearing animals in husbandry systems with higher welfare potential usually incurs extra production costs compared with systems with lower welfare potential. Examples include the costs of reducing stocking densities, providing enrichments and auditing. However, often, there are no extra production advantages for farmers to compensate them for these increased costs. For farming systems with higher welfare standards to be feasible, the marketplace needs to capture the value of animal welfare. This could be from processors, retailers and consumers paying farmers more for food with higher welfare provenance and/or the government channelling public money to support farmers financially so they can achieve high welfare provenance. Agricultural economists have provided some useful insights as to who should pay for this.

The role of government

Consumer surveys across many European countries unequivocally identify animal welfare as a key concern when citizens consider the food they purchase.[24] (Note that price still appears to be the most influential factor when they are actually shopping.) However, free market forces do not automatically capture high welfare in food produce, or even the level of welfare that most consumers would consider acceptable. From a purely economic standpoint, farm animals are 'raw materials' in the human feed system.[25] Their economic value depends on their production. The standard of care they receive is often determined by that needed for producing meat, milk and eggs efficiently (Figure 3-3).

(a) HEALTHYFEET

Mobility score

Category of score	Score	Description of cow behaviour	Suggested actions
Good mobility			
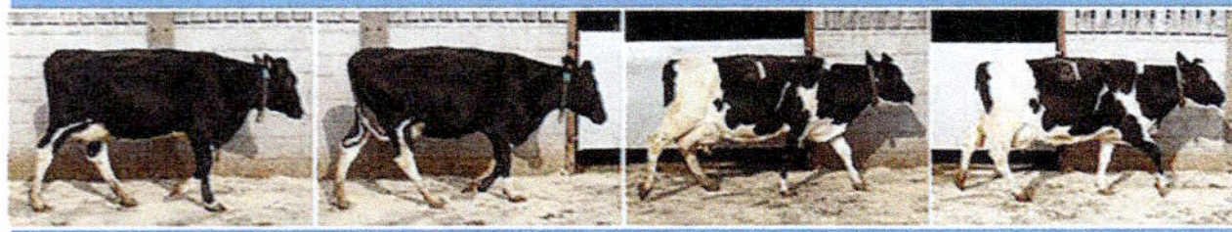	0	• Walks with even weight-bearing and rhythm on all four feet, with a flat back • Long, fluid strides possible	• No action needed • Routine (preventative) foot trimming when/if required • Record mobility at next scoring session
Imperfect mobility			
	1	• Steps uneven (rhythm or weight-bearing) or strides shortened; affected limb or limbs not immediately identifiable	• Could benefit from routine (preventative) foot trimming when/if required • Further observation recommended
Impaired mobility			
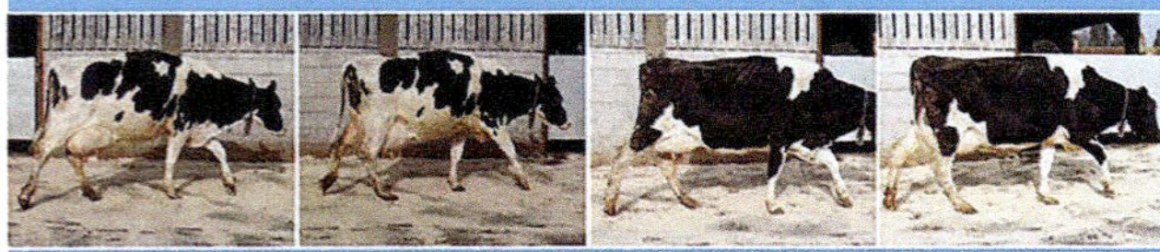	2	• Uneven weight-bearing on a limb that is immediately identifiable and/or obviously shortened strides (usually with an arch to the centre of the back)	• Lame and likely to benefit from treatment • Foot should be lifted to establish the cause of lameness before treatment • Should be attended to as soon as practically possible
Severely impaired mobility			
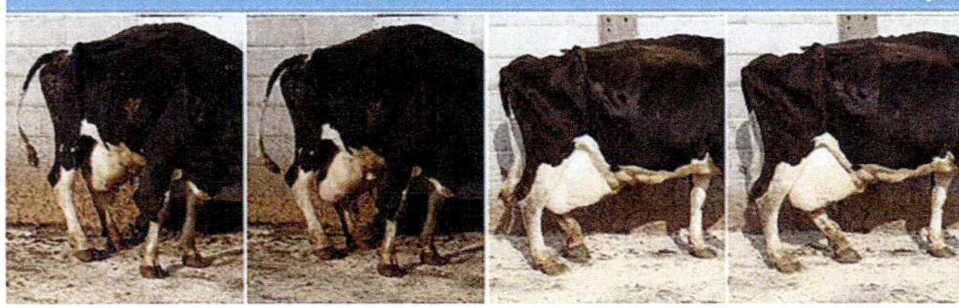	3	• Unable to walk as fast as a brisk human pace (cannot keep up with the healthy herd) • Lame leg easy to identify – limping; may barely stand on lame leg/s; back arched when standing and walking • Very lame	• This cow is very lame and requires urgent attention, nursing and further professional advice • Examine as soon as possible • Cow will benefit from treatment • Cow should not be made to walk far and kept on a straw yard or at grass • In the most severe cases, culling may be the only possible solution

Figure 3-2. Example of welfare scoring systems developed by AHDB to support assessment and action on (a) mobility scoring, and (b) body condition scoring. Reproduced with permission from the Agriculture and Horticulture Development Board.

Continued

(b)

AHDB

Body condition scorecard

What is body condition scoring (BCS)?

BCS was developed around 20 years ago and is widely accepted as a practical means of assessing the impact of negative energy balance in early lactation. Scores range from 1 to 5 in increments of 0.25.

This system concentrates on the accurate determination of scores between 2.0 and 4.0 as these are the most critical for management decisions. Scores outside these values are extreme; those below 2.0 are seriously underconditioned and require immediate attention, as do those at 4.0 and above which are overconditioned and require controlled weight loss.

Scorers using this system will be able to assign BCS consistently and accurately.

When to condition score

The change in body condition score is more important than the absolute value, therefore scoring should be undertaken regularly. A good routine involves scoring:

Stage of lactation	Target BCS
At calving	2.5–3.0
60 days post-calving	2.0–2.5
100 days before drying off	2.5–3.0
At drying off	2.5–3.0

Further information on how to body condition score can be found on the AHDB website, including a short training video, see: ahdb.org.uk/knowledge-library/body-condition-scoring

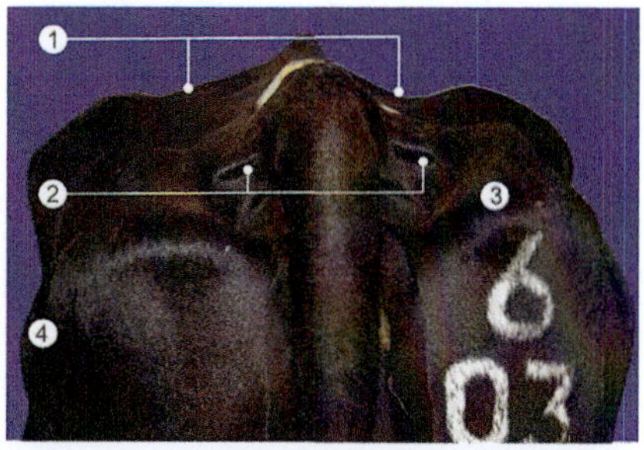

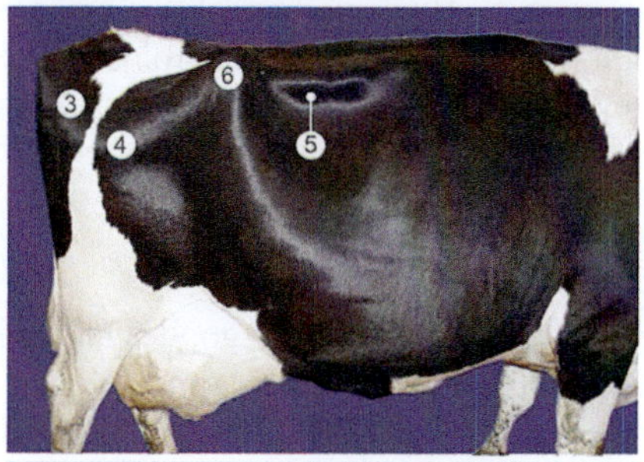

1. Sacral ligament 2. Tailhead ligament
3. Pins 4. Thurl 5. Short ribs 6. Hooks

1 Assess the angle between the hooks and pins.

The first decision you make will divide cows into two groups: those with a BCS less than or equal to 3, and those with a BCS greater than 3.

This decision may be the most difficult one in the BCS process, especiallyif the cow is near a 3.0 or 3.25 BCS.

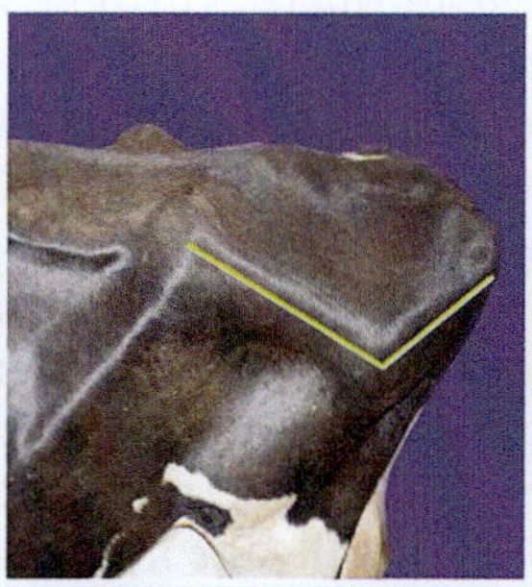

V – angle has a BCS less than or equal to 3. Follow the dark blue steps on page 2, in the left-hand column.

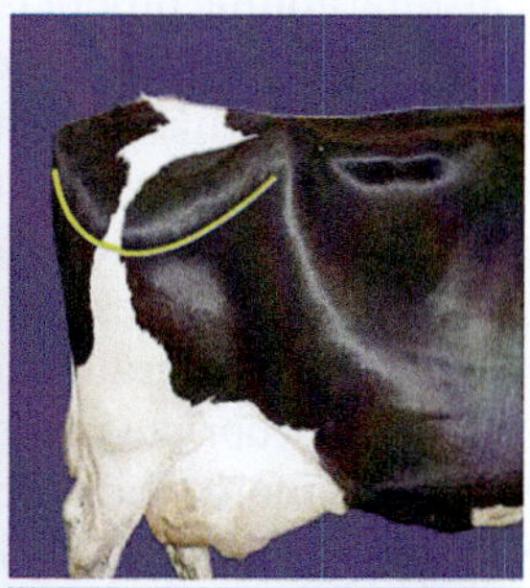

U – angle has a BCS greater than 3. Follow the green steps on page 2, in the right-hand column.

Figure 3-2. Continued.

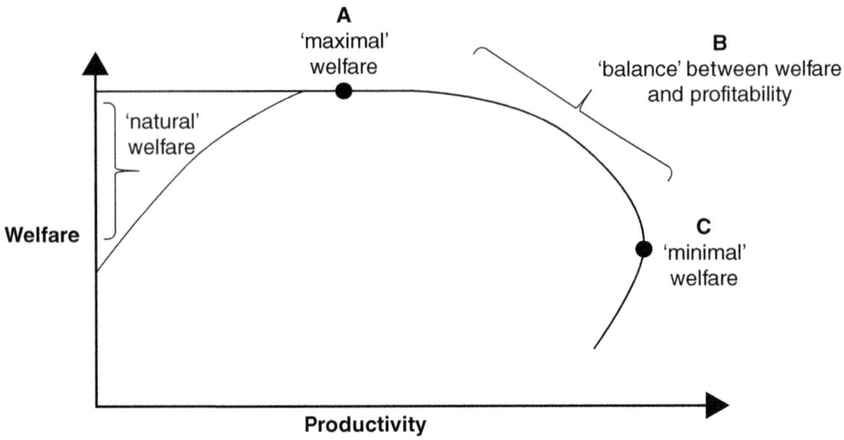

Figure 3-3. The hypothetical impact of increasing productivity on farm animal welfare when standards are determined solely by market forces (adapted from McInerney, 2004).[23] 'Natural' welfare refers to the welfare of free-living, non-domesticated animals. Points A, B and C represent different welfare levels experienced by domesticated animals under human care.

As already discussed, good production does not guarantee good welfare. Generally, 'market forces tend to drive animal welfare standards below socially desirable norms' (McInerney, 2004),[25] as farmers do not automatically recoup the additional costs of higher welfare standards in the prices they are paid.

The welfare of animals matters to many people. It matters to those who consume animal products. It also matters to those who abstain from consuming products due to ethical concerns; economists refer to this as an 'externality', that is, something that impacts persons not linked to the production or consumption of the goods in question. On this basis, animal welfare can be considered a 'public good' or 'societal good'. McInerney (2004) states that where private decisions/market forces alone cannot bring about change, it is the role of government to reflect overall societal preferences and intervene on behalf of citizens.[25]

An appropriate role of government in regulating animal welfare is, firstly, to rigidly enforce the current minimum animal welfare standards (Fig. 3-3, C). Secondly, their role is to help identify the higher levels of animal welfare that some citizens consider necessary (Fig. 3-3, A). At this level, further welfare improvements become a private or minority concern only, and the government is not responsible. Thirdly, they should identify the level of welfare that society and, therefore, government should support, and the incremental improvement that can be made, which usually lies somewhere between the preceding two levels (Fig. 3-3, B). Finally, the government should develop interventions and policies that effectively deliver these shared values. Examples are financial subsidies for farmers who meet appropriate welfare standards. Another less direct form of support is passing laws that mandate the clear labelling of the welfare provenance of food. Ideally, the minimum animal welfare standards mandated by law would be gradually raised to that of an ethical society.

Whilst there is a strong case for the government to intervene in animal welfare, deciding on the best use of resources and the level of acceptable welfare is tricky and subject to ongoing debate. On leaving the EU, the UK government expressed its intention to change financial subsidies for farmers from payments based on acreage to payments for exceeding the legal minimum standards for welfare.[26] Animal welfare is a devolved matter, so the different UK nations can take different approaches.

This chapter has focused on the European approach to animal welfare. However, there are cultural differences regarding animal welfare concerns, which shape the level at which global standards for animal welfare are set by agreement among nations and whether animal welfare is included at all.

The role of the private sector

Consumers can play a key role in raising animal welfare standards through their purchasing decisions, provided they have the option to purchase products with higher welfare provenance. However, such consumer choices are often limited. The food industry may place less importance on animal welfare than wider society. Often, there are commercial incentives for food processors and retailers to source produce with a wide range of animal welfare provenance rather than select only high-welfare produce. Retailers may market products with low and mid-welfare standards as being welfare-friendly to compete with other retailers, believing consumers will not choose the higher welfare products.[27]

Despite this, animal welfare is increasingly recognised as a business risk if not addressed and is often included in corporate social responsibility, alongside issues such as environmental sustainability and worker conditions. Animal welfare is also a business opportunity, as many consumers are likely to be attracted to businesses with a proactive stance. Many UK retailers now have statements and policies on animal welfare. Some work with organisations that support their animal welfare endeavours, such as the Business Benchmark for Animal Welfare (BBFAW)[28] and Food Animal Initiative (FAI).[29] This, in turn, should increase opportunities for farmers and retailers to benefit from selling higher-welfare produce,[30] provided higher payments are shared fairly in the food supply chain.[27] Arguably, farmers are not yet receiving a fair share and transparency is limited.

The animal welfare provenance of food cannot be identified by simply observing the unlabelled food product. Meat, eggs and milk from welfare-friendly systems can look physically identical to those from poorer welfare systems. In addition, consumers often find it difficult to understand the welfare provenance of products from supplier and retailer claims, frequently misinterpreting labels.[27] Their knowledge of farming practices and animal welfare is generally limited.

The diversity of labels probably adds to the confusion. A report found 51 labelling schemes with animal welfare claims across the 27 EU countries,

Switzerland and the UK combined, between 2021 and 2022.[27] An additional 17 labels had an 'organic' dimension where higher welfare claims are generally included alongside health and environmental claims. Some countries had multiple welfare claims and labels, even for a single species, whereas 16 of the 27 EU countries had no national-level labelling scheme with animal welfare claims.

Examples of schemes claiming higher welfare standards are RSPCA Assured in the UK,[31] Label Rouge in France,[32] Animal Partnership[33] in the USA, Beter Leven[34] in the Netherlands, SPCA Certified[35] in New Zealand and RSPCA Approved Farming[36] in Australia.

Many labels on lower-welfare products are easily misinterpreted by consumers, with them wrongly thinking they are more welfare-friendly. Examples from the UK include 'free-range dairy' and 'pasture-fed', terms that the government does not regulate. 'Corn-fed chicken' is a regulated term in the UK, but simply means there must have been a minimum of 35% corn in the diet.[37] In contrast, there are legally protected definitions for various types of poultry farming, such as 'extensive indoor or barn-reared', 'free range', 'traditional free range' and 'free range total freedom'. Some of these systems can benefit poultry welfare, for instance, by providing opportunities for wider behavioural expression.

Some organisations have attempted to help people better understand the plethora of labelling terms. Whilst these are not comprehensive, they are a helpful start. For example, in the UK, the charity Compassion in World Farming (CIWF) provides readily accessible information on the meaning of current commonly used labels and proposals for clearer, simpler labelling.[38] The British Veterinary Association has produced an infographic (Figure 3-4) to help veterinary professionals and wider society choose more welfare-friendly produce.[39]

The need for appropriate labelling

Meaningful labelling that accurately reflects the welfare provenance of animal products is needed to establish trust and confidence among consumers. A common label and standard would be very useful, simplifying the decision making for busy shoppers. EU and UK reports outline strong arguments for making this labelling mandatory and propose possible routes for achieving this.[27]

The extent to which this will develop depends on political priorities. In 2021, the UK government sought opinions from various stakeholders on animal welfare labelling and elicited proposals. The following were put forward.

A simple first step is to label products according to the production method. Clear standardised definitions of farming systems, preferably protected by law, are needed for this to work well. It is then up to the consumer how they rate the welfare value of one system over another.

This approach acknowledges that some husbandry systems have greater potential than others to provide better welfare. For instance, those offering outdoor access, enrichment and lower stocking densities provide animals with greater choice of behavioural expression. Such characteristics align closely with consumers' top concerns, such as prioritising extensive farming over intensive farming/use of cages and outdoor access over indoor confinement.[30]

BVA British Veterinary Association

#ChooseAssured
UK Farm Assurance Schemes Infographic

Below is a reference grid that sets out BVApriorities for farm animal* welfare against what is addressed in the standards of different UK farm assurance schemes. Products may be assured by more than one of these schemes or an assurance scheme not addressed in this graphic. Please check the label of food products carefully.

As part of the #ChooseAssured campaign, BVA is encouraging the veterinary profession and the wider public to #ChooseAssured by purchasing UK animal-derived products that are farm assured. Through the campaign we're raising awareness of the great work of the UK's farm assurance schemes and the crucial work of vets within the schemes to safeguard high animal health and welfare.

*including farmed fish

Please note that this list of the BVA's welfare priorities is not exhaustive and these priorities will be addressed and assessed differently across the different schemes. The level of welfare achieved across the different schemes may vary. For more detailed information about the different standards and requirements used by farm assurance schemes please visit their respective websites.	Farm Assured Welsh Livestock	Lion Eggs Code of Practice	Northern Ireland Beef and Lamb Farm Quality Assurance Scheme	Quality Meat Scotland	Red Tractor	RSPCA Assured	Soil Association
Animals are stunned before slaughter	Assurance does not cover slaughter	Assurance does not cover slaughter	Assurance does not cover slaughter	✓	✓	✓	✓
Veterinary involvement Veterinary professionals are involved in livestock health planning and review	✓	✓	✓	✓	✓	✓	✓
Prohibit environments that substantially reduce behavioural opportunity Enriched cages for laying hens Farrowing crates for sows (pre-birth until weaning)	N/A – Scheme only applies to beef and lamb	Permits enriched cages for laying hens	N/A – Scheme only applies to beef and lamb	Permits farrowing crates for sows (pre-birth until weaning)	Permits farrowing crates for sows (pre-birth until weaning)	✓	✓
Support responsible use of antimicrobials	✓	✓	✓	✓	✓	✓	✓
Animal health and biosecurity Measures to protect animal health and prevent the spread of disease	✓	✓	✓	✓	✓	✓	✓
Lifetime assurance Animals spend their whole lives on an assured farm, livestock transport is assured ie. standards assure the management of health and welfare during transportation and scheme has standards to ensure welfare at slaughter**	Assurance does not cover slaughter	Assurance does not cover slaughter	Assurance does not cover slaughter	✓	Pigs and meat poultry only	All species except dairy – dairy calves can be sourced from non-assured farms	Assurance does not cover transport
Measures to protect the environment ie. guidance on preventing environmental contamination, pollution and minimising waste	✓	✓	✓	✓	✓	Farmed salmon and trout only	✓

**Schemes may address some of these areas even if products are not lifetime assured.

Last reviewed: January 2019, Review date: 2022

Figure 3-4. British Veterinary Association infographic to help citizens choose more welfare-friendly produce.

Pragmatically speaking, it is best to begin with what people are most likely to already understand and care about.

Welfare assessments consisting mainly of animal-based measures (outcomes) should also be used to determine whether the welfare potential of that system is being met, as standards can vary greatly within any system due to differing levels of stockmanship. Assessments must be independent, robust and standardised to provide fair comparisons. Ideally, labelling of all produce will be mandatory rather than voluntary. Otherwise, much produce from the lower welfare systems remains unlabelled and is, therefore, less likely to be rejected by consumers in favour of higher welfare produce.[30]

In EU countries and the UK, a system of production labelling is mandated by law for fresh eggs from laying hens.[40] Free-range, barn and cage systems are clearly defined. This approach is credited with the decline in the use of cages for egg production, from 66% in 2010 down to 28% in 2022.[41] The EU egg labelling scheme started as a voluntary and retailer-led initiative in response to widespread consumer disapproval of battery cages. In 1995, the UK Cooperative Group supermarket labelled shell eggs from hens kept in battery cages as 'intensively produced' before changing the label to 'eggs from caged hens'. Other retailers then adopted similar voluntary labelling. However, this voluntary labelling failed to give consumers clear and transparent information on all eggs. For instance, unlabelled eggs from battery cages continued to be found in some outlets and other terms such as 'farm' and 'country eggs' caused further confusion. Since 2004, shell egg labelling according to production method has been mandatory, with legally defined and protected terms for 'free-range eggs', 'barn eggs', 'caged hens' and 'organic eggs'. Eggs from 'enriched' (furnished) cages must additionally use the label 'eggs from caged hens'.

In contrast, using the legally defined terms for chicken meat production systems in the UK and EU remains voluntary. Legally protected terms are 'extensive indoor/barn reared', 'free range', 'traditional free range' and 'free range – total freedom'. However, most chicken meat production occurs in intensive indoor systems and remains unlabelled.[30]

The UK Animal Welfare Committee has recommended that the mandatory system of production labelling be gradually extended beyond shell eggs, covering eggs used as ingredients. It should also cover meat and dairy produce. The initial focus should be fresh and frozen pig and chicken meat, as these species are commonly consumed, most likely to be kept in barren environments and the production systems are easier to categorise than for other produce.

Practical pain assessments

Pain has been defined by the International Association for the Study of Pain (IASP) as 'an unpleasant sensory and emotional experience associated with, or resembling associated with, actual or potential tissue damage'. Pain is an aspect of animal welfare that is particularly important for veterinary professionals, with animals frequently presenting in pain due to illnesses and injuries and further treatments, such as surgery, inflicting pain despite ultimately benefiting animals. Veterinary professionals are pivotal in guiding animal owners about the quality of life of their animals and when euthanasia may be necessary. To make good decisions, pain must be identified and accurately measured. However, this can be challenging, even for experienced veterinary professionals. Robust pain-scoring tools are useful for improving this skill.

Pain types

Pain can be 'normal' or 'pathological'. Normal, adaptive pain is a protective response. For instance, it alerts an animal to remove the affected body part from the source of pain, thus avoiding or minimising tissue damage. Beyond this, pain serves no useful purpose, becoming 'pathological'. It causes animals to suffer directly and further debilitates them by limiting their ability to find food, shelter and water. It can cause metabolic changes that slow wound healing or lead to immunosuppression.

Normal pain transitions to pathological pain when nociceptors and neurons change in response to tissue damage. In pathological pain, sensitivity to noxious stimuli increases (hyperalgesia), and even non-noxious stimuli such as gentle touch can become painful (allodynia). Pain may be experienced even without any stimulation (spontaneous pain). 'Peripheral sensitisation' is caused by an increased concentration of chemical mediators in damaged tissue, while 'central sensitisation' is caused by altered sensitivity of neuronal pathways due to their repeated stimulation. Central sensitisation occurs mainly in the dorsal horn of the spinal cord and the cerebral cortex and can lead to pain being experienced even after the affected tissues have healed. An example is the chronic, persistent post-operative pain that can occur following inadequate peri-operative pain relief. Many chronic disease conditions, such as cancer or chronic inflammatory conditions, also persistently stimulate the neurons, causing long-term, pathological pain.

Attitudes to pain

Attitudes to animal pain management have significantly improved over the past two decades. Historically, many people used to doubt whether animals and human infants could feel pain, and consequently, pain was often poorly managed. There were concerns that the side effects of analgesics were more harmful than the pain itself. Nowadays, research has provided a much better understanding of pain processes. The negative health and welfare impact of the pain experience is much better acknowledged. Effective analgesics and better ways of measuring pain have been developed. Despite progress, pain management remains suboptimal in many areas and is inconsistent amongst species and contexts. Societal priorities shape research priorities. For example, pain in dogs, cats and horses has been better researched and managed due to their close relationship with humans in many Western countries, making their pain a higher priority for people than that of many farmed and laboratory animals.

Obstacles to the use of analgesics include the costs, availability of licensed drugs and mandatory withdrawal times. These could be partly offset by educating people on the benefits of pain control, both economic and welfare, such as better animal recovery due to increased mobility and access to food. However, some types of pain, like lameness in cattle, require significant upfront investments in breeding, housing and management, which can be prohibitive.

Researching animal pain

Researching pain is quite similar to researching other aspects of animal welfare, building on our understanding of pain in humans and measuring physiological, physical and behavioural changes in animals (see Chapter 2).

Human pain experiences have been useful for understanding the likely pain experiences of animals and their impact on welfare. Descriptions by humans indicate that the pain experience is highly variable. They report sensations such as stabbing, throbbing, shooting and crushing, and accompanying feelings such as anxiety, fear, depression and nausea. Pain is influenced by many factors, such as a person's age, sex, temperament, previous experiences, location and type of pain stimulus, familiarity with the environment and concurrent disease. It is reasonable to assume there are similarly diverse experiences of pain amongst non-human animals. In addition, the animal kingdom is incredibly diverse, with a wide range of anatomies and physiologies, and some animals may additionally experience pain in ways that humans cannot envisage. This makes studying pain and creating pain-scoring instruments in animals even more challenging than in humans.

Research in humans and animals shows that the pain experience is linked to detectable and quantifiable physical, physiological and behavioural changes. Some have been extensively researched and incorporated into practical pain-scoring systems that veterinary professionals can use. Developing reliable research techniques is a slow and meticulous process. To date, acute pain has been more extensively researched than chronic pain, with the management of post-operative pain, post-trauma pain and musculoskeletal pain being prioritised over pain due to inflammatory conditions like osteoarthritis, neuropathic pain or cancer pain. Some aspects of farm animal pain have received considerable research attention in animal welfare science. For example, concerns about the pain caused to farm animals by 'mutilations', such as castrating and tail docking in lambs, often without anaesthetics or analgesics, have led to research prioritising this area.

One research challenge is that it can be difficult to design studies that distinguish between nociception (stimulation of pain receptors) and the actual conscious pain experience in the animal. For instance, research has shown that physiological measures, such as heart rate, blood pressure and endocrine responses, can alter in animals during anaesthesia or handling alone, in the absence of painful procedures. Therefore, study designs must account for this, including control groups where animals undergo only handling and 'sham' procedures or only anaesthesia.

Another challenge is addressing ethical considerations where studies require animals to be in pain. For example, helpful study designs include comparing observations of animals undergoing procedures with and without analgesics, comparing animals with and without injuries, comparing signs before and after treatment with anaesthetics and analgesics and observing the amounts of analgesics that animals choose to consume. However, it is essential to consider the ethics of such studies, for example, ensuring any pain and suffering stays within acceptable limits and is sufficiently justified by the potential benefits for future animals (see Chapter 16). For example, some studies must be conducted under anaesthesia on ethical grounds, which complicates the interpretation of results.

Development of pain-scoring instruments

Desirable features of pain scoring instruments are similar to those recommended for general welfare assessment protocols (see Chapter 3). They include using animal-based measures underpinned by expert opinion and science, which are valid, repeatable and feasible. Behavioural changes are the mainstay of pain-scoring instruments, as they are easily observed in real-time. They include behavioural observations of the undisturbed animal and in response to handling. Where scorers deal with individual animals or small groups of animals that are confined and easily accessible, physiological measures such as heart rate and respiratory rate are often practical to perform. Clear criteria and training are required to ensure scores are properly used and applied in a consistent manner across all veterinary team members.

Knowledge of species-specific behaviour is important both when developing and using scores. For example, prey species such as sheep or rodents often show subtle signs of pain and will further mask these when handled or if they know they are being watched. The pain behaviours of nocturnal species, such as mice, may be more obvious during the night when they are more active but no one is watching. Rodents in acute pain emit ultrasonic vocalisations that humans cannot hear. When using scores, it is also important to know the baseline for the individual animal. For this to be accurate, time must be spent with the patient, talking to the owner and keeping good records.

Structures of pain-scoring instruments

Pain is generally measured using either 'unidimensional' or 'multidimensional' composite scales. Unidimensional scales have only one ('uni') dimension;

a single number line can represent them, and the pain intensity is assigned a single score. They are often simple descriptive scales (SDS), visual analogue scales (VAS) and numerical rating scales (NRS). Simple descriptive scales usually provide 3–5 severity grades, each defined by a short description (Table 3-2). With visual analogue scales, the patient is generally observed in the kennel with no interaction. A mark is made at the position on a line where the pain appears to lie, with one extremity of the scale representing no pain and the other extremity representing the most extreme pain. Numerical rating scales are similar to visual analogue scales, but a number on a scale is chosen instead of a location on the line. These unidimensional scales are straightforward and can provide helpful information about acute pain. However, they are very subjective and only assess a limited part of the pain experienced.

Multidimensional pain scales capture additional aspects ('dimensions') of the pain experience. They measure various factors, such as pain intensity, type and location, mood changes, anxiety levels, depression symptoms and everyday functioning. The animal is observed moving and interacting with humans, as well as lying undisturbed. Responses to palpation are assessed along with mental state and any vocalisation. Often, objective physiological data, such as heart rate, altered respiratory rate and depth, temperature and pupil size are included. The observer scores each aspect according to the descriptors provided. Many multidimensional scales are also 'composite' scales; that is, the scores for the different aspects can be combined into a single summary measure. Multidimensional scales provide a more detailed measure of the pain experience and are certainly a better measure of chronic pain than a single-dimensional scale (Figure 3-5). However, the main limitation is the time to complete. Facial grimace scales have been developed for some species by observing and mapping facial expressions to noxious stimuli. Measures include the amount of orbital tightening and position of the nose, ears, cheeks and whiskers. Facial grimace scales can be used in isolation or as part of multidimensional scales. When choosing an appropriate pain scale, a balance must be struck between obtaining accurate information on the pain experience and practicality.

Table 3-2. Example of a simple descriptive scale in cats.[42]

Score	Description
0	No pain, no overt signs of discomfort and no resentment to firm pressure
1	Some pain, no overt signs of discomfort, but resentment to firm pressure
2	Moderate pain, some overt signs of discomfort, which are made worse by firm pressure
3	Severe pain, obvious signs of persistent discomfort, which are made worse by firm pressure

Glasgow Feline Composite Measure Pain Scale: CMPS - Feline

Choose the most appropriate expression from each section and total the scores to calculate the pain score for the cat. If more than one expression applies choose the higher score

LOOK AT THE CAT IN ITS CAGE:

Is it?

Question 1

Silent / purring / meowing	0
Crying/growling / groaning	1

Question 2

Relaxed	0
Licking lips	1
Restless/cowering at back of cage	2
Tense/crouched	3
Rigid/hunched	4

Question 3

Ignoring any wound or painful area	0
Attention to wound	1

Question 4

a) Look at the following caricatures. Circle the drawing which best depicts the cat's ear position?

0 1 2

b) Look at the shape of the muzzle in the following caricatures. Circle the drawing which appears most like that of the cat?

0 1 2

Figure 3-5. Example of a multi-dimensional composite score for cats. Reid J, Nolan AM, Hughes JML, Lascelles D, Pawson P, Scott EM. Development of the short-form Glasgow Composite Measure Pain Scale (CMPS-SF) and derivation of an analgesic intervention score. Animal Welfare. 2007;16(S):97–104. © Zoetis. For more information, visit www.newmetrica.com.

Continued

APPROACH THE CAGE, CALL THE CAT BY NAME & STROKE ALONG ITS BACK FROM HEAD TO TAIL

Question 5

Does it?	
Respond to stroking	0
Is it?	
Unresponsive	1
Aggressive	2

IF IT HAS A WOUND OR PAINFUL AREA, APPLY GENTLE PRESSURE 5 CM AROUND THE SITE. IN THE ABSENCE OF ANY PAINFUL AREA APPLY SIMILAR PRESSURE AROUND THE HIND LEG ABOVE THE KNEE

Question 6

Does it?	
Do nothing	0
Swish tail/flatten ears	1
Cry/hiss	2
Growl	3
Bite/lash out	4

Question 7

General impression

Is the cat?	
Happy and content	0
Disinterested/quiet	1
Anxious/fearful	2
Dull	3
Depressed/grumpy	4

Pain Score .../20

Figure 3-5. Continued.

Validation of pain-scoring instruments

It is preferable to choose a score that has been validated, that is, it has been checked that the measures used in the scoring system are valid, repeatable and feasible. Levels of validation vary amongst currently available scores. Additionally, it is important to remember that a pain scale developed for one animal strain in a particular context may not apply well to different strains and contexts.

For many farmed animals, dogs and cats, there are sufficiently validated instruments to choose from. A thorough scientific review of the validity of pain-scoring instruments for farm animals[43] identified 20 pain-scoring tools for three species (sheep, cattle and pigs) that comprised behaviour and/or facial expression scores. It found moderate to strong evidence underpinning the validity of seven of the scores. There was little evidence supporting the validity of the remaining scores, but this was because the validity had not yet been tested, as opposed to scientific evidence throwing doubt on the scores' usefulness. As for general welfare assessment protocols, developing pain scores is an iterative process – they improve with usage, feedback and further research developments.

Pain: further resources

The Animal Pain website (https://animalpain.org) is a valuable resource developed by the Pain and Quality of Life in Animals project at the Unesp-Botucatu campus in Brazil. It contains pain scales for dogs, cats, cattle, pigs, horses, sheep, rabbits, donkeys, rats and mice. The website provides in-depth information about each pain scale, including its criteria, structure and relevant scientific articles that underpin the scores. Additionally, the site provides video cases that allow you to test your own usage of scores.

Also, the WSAVA Global Pain Guidelines[44] and supporting information are an excellent resource for dogs and cats. Ongoing initiatives, such as VetMetrica,[45] are making pain-assessment tools more user-friendly and readily available on electronic devices. They are also using data to benchmark scores and help users interpret the scores.

4 Ethics

Abstract

This chapter sets out the benefits of engaging further with the discipline of ethics. It delves further into what ethics actually is, including the relationship with professional codes of conduct, and what engaging in ethics can and cannot achieve. Some key ethical theories (purpose, principle and consequence) and frameworks (the 3Rs and the ethical matrix) are explained. These are then incorporated into a suggested structure for navigating ethical discussions, introducing additional useful skills such as evaluating arguments. This approach is briefly illustrated using an example of 'street dog' control. By further developing these skills, veterinary professionals will be better equipped to advocate for animals when discussing controversial issues and to effect practical change.

Introduction

Decisions with an ethical dimension impact the well-being of others, whether humans or animals, and also our own well-being. In everyday conversations, the words ethics and morals are used interchangeably and taken to mean right or wrong beliefs and conduct. However, ethics is so much more than this. Ethics is a branch of philosophy that analyses why we consider something right or wrong. The field of applied ethics provides tools and techniques to help people structure their reflections and make better decisions.

Ethics is probably needed now more than ever, given the fast-paced society we now live in, with quick responses on social media. Communication in the modern world does not lend itself to thorough, nuanced discussions. This chapter aims to convince veterinary professionals that the subject area of ethics can be valuable in practical decision making in the real world and should not be relegated to academia, to be parked as soon as they have jumped through the various academic hoops required to achieve qualifications.

DOI: 10.1079/9781786391582.0004

Ethical challenges facing veterinary professionals

Veterinary professionals face numerous ethical challenges. They often deal with controversial issues. These can be local decisions. For instance, they must balance their interests in maintaining a viable business with those of their animal patients and owners. Veterinary professionals can also be pivotal in questioning the overall context of animal usage in society – the 'bigger picture' – and continually striving to improve societal relations with animals. For example, there are many welfare considerations underpinning activities such as horse racing, farming and keeping animals as 'pets', all activities that society has widely accepted based on its current knowledge. However, if citizens become better informed, they may be more likely to support change and stronger safeguards for animal welfare.

Many ethical issues are unique to the veterinary field. For example, unlike in human medicine, the moral status of the patient is not necessarily agreed upon in all cases. In many countries, human life is considered sacrosanct, and protecting it is not questioned, whereas animal life is not. They have a utility value to us and are killed as we consider 'necessary'.

Veterinary professionals are in a strong position to negotiate on behalf of animals, ensuring that their interests are included, accurately represented and respected in decisions that concern them. Whilst welfare knowledge and science can inform decisions, they alone cannot give answers in the real world. By engaging with the discipline of ethics, veterinary professionals are better placed to know, justify and articulate their ethical positions and challenge those of others. Society expects to be able to trust veterinary professionals in matters concerning animal health and welfare, particularly veterinarians and veterinary nurses, who are governed by their respective professional codes of conduct. Welfare legislation may reflect this in allowing certain activities that pose welfare risks, provided there is active veterinary involvement to mitigate.

Bernard Rollins, a veterinary ethicist, clearly stated the role of ethics. Veterinary medicine *'should be continually improving itself through the twin disciplines of science and philosophy. Science gives evidence for the best treatment, and philosophy the reasons for the moral choices made ... Just as it takes training and practice and a conceptual map of the medical possibilities to learn different diagnoses of disease, so too it takes training and practice to dissect out all the ethical nuances of many complex situations'*. He goes on to say that without this, *'we may know something is problematic but have trouble saying what the problem is'*, and that *'we tend to react to one obvious component and ignore all others'*.[46]

Engaging in ethics is useful as it can:[47]

- Provide us with tools for thinking about moral issues.
- Provide us with a structure for working with others, especially those with different views, promoting dialogue and proper discussion.
- Make us challenge our assumptions and the weight we place on different values. Our assumptions are shaped by our individual background, culture, education and personal experience.

- Pinpoint the areas of agreement and disagreement, clarifying issues and eliminating confusion.
- Focus us away from self-interest, making us consider someone or something other than ourselves.
- Help us develop, interpret and implement our professional regulatory bodies' ethical codes and rules.

However, ethics also has its limitations and engaging with it cannot necessarily:[47]

- Give one right answer – ethics often gives several different answers, and life has much moral ambiguity. The complexity of possible actions and their consequences often leads to multiple interpretations, making it difficult to determine the best answer.
- Change fundamental assumptions and weighting of issues that are deeply embedded.
- Make it easy to resolve ethical issues.

Ethics and professional Codes of Conduct

In countries where veterinary professionals are well regulated, expected standards of behaviour are often set out as professional guidance in the form of a Code of Conduct. For example, in the UK, there is the Royal College of Veterinary Surgeons' Code of Professional Conduct for Veterinary Surgeons and the Code of Professional Conduct for Veterinary Nurses.[48] Such Codes are often referred to as veterinary ethics. However, veterinary ethics involves so much more. Veterinary ethics involves critically questioning the morals underpinning the principles in professional codes and guidance, challenging the status quo, seeking justification and examining what ought to be rather than simply accepting what currently is. You may see this style of ethics referred to as 'normative ethics'. If anything, professional codes of conduct and associated guidance are the outcome or by-product of ethical discussion about the role and purpose of the veterinary profession. As with morals in other walks of life, they are subject to change, mirroring ongoing professional debate and social and material influences. Improved skills in ethical analysis should support better discussions and decisions.

Societal morality and animals

Different societies have differing assumptions about animals and how people should interact with them. For example, in many Western countries, many people believe that animals:

- Have some moral status due to their sentience.
- When in our care, should be looked after.

- Should not be killed unless 'necessary' (here, necessity includes meeting our needs or wants for animal products such as food).
- Should have a life worth living and a humane death.
- Quality of life is important, rather than quantity of life.
- Are the property of humans.

These societal morals are the 'taken for granted notions' of how things are in life in Western societies. They are the 'backdrop' against which we perform our veterinary roles. However, this is not the only way a society can or should function.

Travelling and working in other cultures can help us become more aware of things we often take for granted, which is important if we want to make a practical contribution to the welfare of animals in other societies, or even simply to avoid doing harm. It is also valuable for making us aware of the various taken-for-granted notions of how things are in our own societies.

Proper ethical consideration of issues often challenges this 'status quo'. For example, ethical analysis of the Western approach to animals reveals many inconsistencies in viewpoints. How animals are treated often depends on their use or nuisance to humans, their emotional appeal and their relationship with humans rather than something more morally relevant, such as their level of sentience. For example, often, people in the UK abhor the poisoning of 'street dogs' (unowned, free-living dogs) yet are indifferent to inflicting an inhumane death on rats that we label as 'vermin' or 'pests', despite these species sharing similar sentience. The philosopher, Peter Singer, would describe this as an example of 'speciesism'. The potential harm of rats to humans is often used to justify their eradication, even by methods that can cause pain, distress and suffering. For example, in the UK, the use of glue traps for controlling wild rodent populations is permitted and has only recently become subject to greater legal safeguards (Glue Trap (Offences) Act 2022). Though the Western approach to animals is increasingly dominant, it is only one approach. For example, in India, rats may be revered in some areas due to their association with a deity (Figure 4-1).

Another cultural norm in many societies is for people to see themselves as having a greater duty to those people and animals with whom they have particular bonds.[49] Often, people form a strong bond with companion animals, such as dogs, cats and horses. Young and injured animals often elicit higher emotional responses in humans, as do 'cute' animals or animals with which humans can empathise. As a result, these animals receive greater moral concern than other species of similar sentience, with animal welfare charities for dogs, cats, horses and donkeys often eliciting more donations. Indeed, people can be more adversely affected by the suffering of animals they care for or are emotionally attached to, and this is a consideration when balancing the harms and benefits of different courses of action. However, veterinary professionals should probably challenge this thinking to improve animal welfare more widely. For example, we may work with a client to do our best for the welfare of a beloved 'family member', whether that be a pet dog, cat or horse, but also

Figure 4-1. The Karni Mata Temple in Rajasthan, India, is populated by revered holy rats, often seen drinking from bowls of milk left by devotees [left]. Contrast with two rats trapped on a glue trap [right], which is used to kill them and control their numbers. Rats and mice often remain attached for considerable time periods on glue traps before being killed or dying from trauma or starvation.

gradually educate owners on the usage of other animals in society and routes to better cater for their welfare. Perhaps some of their compassion can extend to these areas. Be mindful that we all use animals indirectly in society. For example, every medicine used to help a companion animal will have impacted the welfare of laboratory animals, and many dog and cat foods contain products derived from farmed animals.

Various factors shape societal morals. Societies comprise individuals who hold varying views based on their personal morals. They share many morals that often become supported by laws and regulations, although not everything considered ethical is supported by law, and vice versa. Individuals learn their morals from various sources, such as the home and different institutions. The term 'institutions' usually refers to organisations such as schools, banks, hospitals, prisons and broadcasting corporations that we encounter daily. However, in some disciplines, institutions are understood as a 'social construct' – a way of thinking about the formal and informal rules ('norms') that shape social, political and economic relations and ultimately shape human behaviour.[50] Animal welfare can be considered an institution in this sense. Understanding power dynamics is key to understanding institutions. Power dynamics can be overt, subtle or even hidden. They determine which competing interests are favoured and whose opinions are more widely heard. For instance, resources and finance strongly influence whether opinions are articulated persuasively and how well they reach people. Some individuals and groups have greater power than others to promote certain norms and rules. Understanding power dynamics and who controls the narrative can help us understand why situations are as they are and how to focus our efforts to bring about change. It is beyond the scope of this book to discuss different institutions in detail, but below are two examples of powerful institutions that are influential in animal welfare: religion and media.

Religion and animals

Religion can be defined as a system of beliefs that typically involves the worship of one or more gods and provides ethical guidance through religious leaders, often referring to religious texts that offer advice on living a good life. Religions such as Christianity, Islam, Hinduism, Sikhism, Judaism and Buddhism have been the primary moral codes for many societies and have a significant influence on how different societies view and interact with animals.

The extent to which religious texts and practices refer to animals varies, and their interpretation has varied over time and across cultures. Eastern and Western religions have shaped their respective societies and held very different views on the moral status of animals. In many Eastern religions, all life is considered sacred. For example, Jainism, Hinduism, Sikhism and Buddhism all accept some form of the doctrine of 'ahimsa', which advocates non-injury to all living things and reverence for all life. These traditions share beliefs in reincarnation, with Jains and Buddhists emphasising the interconnectedness of all living things and recommending vegetarianism. In contrast, Christianity's worldview is that humans are predestined to use natural resources, including animals, for their purposes. 'Man was made to rule (have dominion) over the fish in the sea and the birds in the sky, over the livestock and all the wild animals, and over all creatures that move along the ground (Genesis 1:26).' Religions have provided the moral underpinning of many societies, providing the backdrop of moral assumptions against which different philosophers have developed their views of animals, for example, Aquinus, Descartes or Kant. As society has become increasingly secular, with a growing proportion of atheists, there has been a need for alternative moral frameworks or decision-making processes. In Western society, the utilitarian approach to animals remains dominant, but philosophers are questioning certain aspects of it. Additionally, religion has evolved in response to societal pressures, changing its interpretations accordingly. For example, in some societies, Christianity now advocates for stewardship (taking care) of animals over dominion (dominating them).

The power of the institution of religion is reflected in laws permitting non-stun slaughter for religious reasons in the UK, whilst generally banning its use on welfare grounds for other reasons. The legal requirement to stun animals is unique in applying differently to citizens depending on their religious beliefs, rather than to all citizens based on the animal welfare impact of the killing technique (see Chapter 17).

Media and animals

Media can reach very large audiences, using tools such as broadcasting, publishing and the internet, and the messages are often compelling. Where the media uses informed, accurate materials created by credible individuals and organisations, there is scope to advocate for animals. However, media often unintentionally conveys messages that can harm welfare. For instance, the film 'Patrick' inadvertently promoted the popularity of pug dogs, a breed known to experience welfare harms due to their extreme conformation.

Unfortunately, the filmmakers had not considered this issue, despite position statements of credible organisations such as the British Veterinary Association (BVA), which recommended not using images of such breeds. The message of the BVA and other welfare organisations was 'drowned out' by much more powerful media activities. The development of social media, such as TikTok, YouTube and Instagram, is likely to have magnified such problems by creating largely unregulated spaces where anyone – whether influencers, celebrities or members of the public – can post content involving animals. Such materials can have a far-reaching impact, potentially viewed by millions of people of all ages and backgrounds. Anyone can self-style themselves as an expert, readily spreading misinformation on animal care, behaviour and training. There is widespread creation and posting of video footage of animals, intended to be entertaining and create viral content. For example, there is significant demand for 'funny cat' videos. Relevant animal welfare research is still in its early stages, but these videos have already been found to place animals in situations that harm their welfare, with those creating and viewing the content often oblivious to their suffering.[51,52] Such footage may normalise poor practices and encourage viewers to mimic what they see. Other examples of harmful content include that which encourages and normalises impulsive pet purchases.

Ethical theories

People vary in their views. Philosophers have developed ethical theories that can help us understand our views and those of others. In turn, these theories can guide decision making.

Three key ethical theories are:

- Ethics of purpose.
- Ethics of principles.
- Ethics of consequence.[47]

In academic texts, you may see ethics of purpose referred to as 'virtue ethics' and ethics of principle as 'deontology'. 'Utilitarianism' is the most commonly offered interpretation of ethics of consequence.

Ethics of purpose

Ethics of purpose focuses on the integrity of the person. It includes the 'virtue theory' of ethics, the idea that right action will automatically stem from virtuous personal qualities such as compassion, empathy, honesty, selflessness, conscientiousness and putting others first. When analysing or creating a proposed action according to this theory, identify the person or body proposing it and then identify their purposes: What kind of person or organisation should they be? What personal attributes should they have? What should they try to achieve? An example would be the proposal that veterinary professionals should be knowledgeable, empathetic and conscientious to the degree that they automatically put the animal's interests ahead of any financial self-interest.

Ethics of principle

Ethics of principle focuses on the rules underpinning a proposed course of action. Examples of general rules are never to kill a healthy animal or never perform a procedure that causes an animal to suffer unless that individual benefits from the procedure, regardless of the overall benefits for other animals and humans. When analysing or creating a proposed action according to this principle, consider whether it respects the animal as an individual and the animal's sentience and interest in avoiding suffering. Also, question whether the proposed action is fair. Ethics of principle requires moral consistency by treating others in similar situations the same way, rather than making exceptions. This is often referred to as the 'principle of justice'. An example of moral inconsistency is the preferential treatment of dogs solely because they are pedigree, young or cute, despite all dogs having a similar level of sentience.

Ethics of consequence

This ethical theory focuses on the consequences of a proposed action. It lists the stakeholders, that is, anyone or any animal with something to lose or gain from the proposed action, and examines its likely impact on them. Utilitarianism aims to deliver the 'greatest good for the greatest number', i.e. you should do the action(s) where the overall benefits most outweigh the harms. However, it can be challenging to weigh up the different positive and negative outcomes, plus not all outcomes can be predicted. Additionally, stakeholders may have varying levels of representation or understanding. For example, the more powerful commercial organisations can often pay to represent their vested interests, whereas animals cannot represent themselves. Animals require well-informed advocacy for their interests to be heard and to compete against other interests.

Often, analyses based on the ethics of purpose, principle and consequence lead to different and conflicting answers. Ethical texts tend to highlight the difference between theories. As a result, there is a danger that they will be seen as having limited value when making decisions in the real world. However, this overlooks the value to be found in exploring an ethical issue from all angles, broadening the horizons of participants and increasing the number of resources used to inform a decision, thereby strengthening the decision. We would like to stress that **the best decisions will have thoroughly considered the issues using all three approaches – purpose, principle and consequences.**[47]

Ethical frameworks

Ethical frameworks combine ethical theories into a more practical working structure. They act as prompts, expanding our thinking about the various possibilities for action. We can then consider and accept or discount these possibilities, giving justified reasons. They are useful for structuring discussions with others or working through an issue alone. Two commonly used frameworks are the ethical matrix and the 3Rs.

The ethical matrix

The ethical matrix was developed by adapting the 'four principles of medical ethics'[53] – principles created to guide the interactions of human doctors with their patients. Here, the term principle means a general guiding statement, not to be confused with 'ethics of principle', described above.

The matrix uses three broad principles. These are:

- Choice: respecting the choices of the stakeholders.
- Well-being: doing good and avoiding harm.
- Fairness: being morally consistent, for instance, treating like cases similarly.

Here are examples of the various stages and questions you would consider when using an ethical matrix:

- Identify all the stakeholders, that is, anyone with something to lose or gain from the decision.
- Identify the interests of each stakeholder according to the ethical norm.
  - What would doing good and avoiding harm on their behalf involve?
  - What would 'being fair' to them involve?
  - What would 'respecting their individual choices' involve?
- Assess whether the proposed course of action respects choice, well-being and fairness for each stakeholder.
- Are we respecting or infringing any 'principles' for any of the stakeholders?
- How much are we respecting or infringing?
- What are our and other people's priorities and why?
- Why are we respecting or harming each 'principle'?
- Does this matter? Why does it matter? And so on ...?

This process lays the foundation for further debate and research on the controversial issue. Consensus can be reached by proposing alternatives or modifying the current proposal. Tables are often used to present the key points so they can be easily visualised, helping participants keep track of the discussion and ensuring that all important areas are covered (Table 4-1). Different people have different approaches to using the ethical matrix. Some find this framework a

Table 4-1. Application of the ethical matrix to the use of animals in scientific procedures, regulatory toxicology and drug testing (adapted from Webster (2014)).[54] When using an ethical matrix, stakeholders can be expanded upon according to context, including farmers, producers, consumers, citizens, retailers and the environment.

	Principle		
Stakeholders	Well-being	Choice	Fairness
Animal	Respect for welfare.	Behavioural freedom. Avoiding suffering.	Seen as having inherent value, no lower than species of similar sentience that humans hold in high regard. Certain harms are considered so severe that they are not allowed and cannot be overridden by a cost–benefit analysis.
Animal care staff	Pride in their work through providing a high standard of care for the animal. Security at work.	Control and input into decisions, such as husbandry and euthanasia.	Input into animal welfare policy.
Citizens/human society at large	Improved health, e.g. safe food. Product safety, e.g. safe medicines.	Freedom to choose various therapies and products.	Informed about the harms to animals used in tests.
Regulators of products and of animal experiments	Meeting health and safety responsibilities to society. Meeting responsibilities to animals through minimising harm.	An open-minded approach to new developments, e.g. testing methods.	Respect for animal welfare enshrined in legislation and codes of practice, creating a level playing field for competition and establishing a minimum standard below which businesses must not fall.
Scientists	Informed and sympathetic to regulations and procedures. Financial reward.	An open-minded approach to new developments, such as new testing methods.	Fair interpretation of the legislation. Proper and consistent application of the 3Rs.[55]

useful starting point for ideas, while others may prefer to jot down all their thoughts about the issues at hand before determining where each aspect fits in the matrix and filling in any gaps.

The 3Rs

The concept of the 3Rs first appeared in the book 'The Principle of Humane Experimental Technique' by two British scientists (Russel and Burch) in 1959.[55] They developed three Rs – replace, reduce and refine – as a decision-making tool for improving the welfare of laboratory animals. They argued that humane animal use was an integral part of valid science, as the physical and mental stresses placed on laboratory animals alter their physiological variables in a way that skews and distorts scientific results.

The 3Rs is an example of a utilitarian framework. It can be used to identify how the harms of a chosen course of action can be minimised, capturing the number of animals impacted as well as the intensity and duration of harm experienced by individual animals.

- **Replace**: Aim to replace animals with non-sentient alternatives. Examples include studies *in vitro*.
- **Reduce**: Aim to reduce the total number of animals used. Examples include efficient planning and more precise use of statistics.
- **Refine**: Aim to minimise pain and distress and optimise welfare. Examples include altering procedures and experimental techniques so they are less harmful, proper use of analgesics and anaesthetics, proper handling techniques, using fully-trained animal carers and providing environmental enrichment.

The 3Rs are also valuable for resolving issues in veterinary medicine and other animal-related fields, finding the 'best way' forward. A helpful resource is an article by Emily Gascoigne and others describing its application to castrating and tail docking lambs.[56]

The 3Rs framework has become widely accepted in the research community across most European countries; indeed, it underpins EU law protecting animals used for scientific purposes. However, it must be emphasised that there is much more to ethics than the 3Rs. There needs to be ethical scrutiny of whether a proposed course of action can be justified in the first place, with the 3Rs simply minimising harms once the course of action has been given the go-ahead.

Navigating ethical discussions

Using ethical theories and frameworks pushes us to consider the values underpinning our decisions and to examine issues from a wider range of perspectives. This should deliver better decisions and proposals for action. From the veterinary professional's perspective, a good decision can be regarded as one that advocates for animals, with the animal's interests included, properly represented and respected. However, exploring ethical values is only one part of the decision-making process. Additional skills are needed.

The authors have found the Ethical Process developed by Marvin T. Brown[47] particularly useful. The process is outlined below and illustrated using population control of 'street dogs' as an example.

Steps in the Ethical Process

1. Agree to engage in dialogue.
2. Assess whether an argument is being put forward.
3. Reconstruct the argument so it is clearly expressed.
4. Evaluate the argument.
    a. Evaluate the logic.
    b. Assess the facts.
    c. Assess the values.
    d. Assess the assumptions.
5. Modify the proposal accordingly.

The 'ethical process' in action

Scenario

'Street dogs' are free-living dogs commonly found in human settlements in many countries (Figure 4-2). Controlling their population is controversial, with different stakeholders advocating different methods, from killing the dogs through to neutering and vaccinating them, and possibly rehoming them. Animal welfare charities often get involved in neutering and vaccinating these dogs against rabies and releasing them where they were found, arguing that 'street dogs should not be culled because it does not work'. What might your position be as a veterinary professional working with animal charities and other organisations involved in street dog control?

1. Agree to engage in dialogue.

Engaging in dialogue first requires a willingness to properly listen to others, which can be particularly challenging where there are strong opinions. Disagreements can be beneficial, provided they are used wisely. Brown states that '*the quality of our decisions depends on the quality of resources we use to make them. No one person has all the good resources. When people disagree about what to do, they provide an occasion for investigating the reasons behind different views. This, in turn, uncovers more resources – strong evidence, value judgments and assumptions – giving individuals and groups more resources than they had before.*

Figure 4-2. 'Street dogs' in India.

There is an opportunity to learn from others. Although having more resources does not always mean the end of disagreements, it does assure everyone that the decision rests on consideration of the best resources'.[47]

However, disagreements also carry risks: favouring argumentative types over others, stifling participation, threatening cooperation, delaying action and creating a debating game of winners and losers. This is why a logical, step-by-step process can be helpful when working as a group. It can minimise the risks and maximise the benefits.

2. Assess whether an argument is being put forward.

An argument is a set of reasons that support a conclusion and is used to persuade us to believe or do something.[57] In this example, an argument has been made, as there is a clear reason for the proposal: 'Street dogs should not be culled *because* it does not work'.

3. Reconstruct the argument so it is clearly expressed.

Ensure clear definitions are given. Address what we mean by a 'street dog'. Are they free-living dogs that associate with certain households, abandoned dogs, feral dogs, or are we grouping several categories of dogs? Define what we mean by 'culling'. Are we referring to humane killing methods, inhumane killing methods, killing of healthy dogs and so on? Establish what we mean when we say culling is 'not working'? Do we mean it doesn't control rabies incidence in dogs and humans? Or that it doesn't remove the perceived nuisance of groups of dogs in parts of the city?

For this example, 'culling' is taken to mean any 'killing'. 'Not working' is taken to mean 'does not control rabies incidence in dogs and humans'.

People often do not express themselves clearly, and they regularly leave parts of their reasoning implicit or hidden. Techniques in argument reconstruction help identify these reasons.

We need to add a reason if we are to logically get from Reason 1 to the conclusion that we should not kill street dogs.

Reason 1: 'Killing street dogs does not work'
Reason 2: 'We should not do things that do not work'
....
Conclusion: 'We should not kill street dogs'

4. Evaluate the argument according to its logic, supporting facts and evidence, underpinning values and assumptions.

a. Evaluate the logic.

Do the reasons lead logically to the conclusion? Are there mistakes in reasoning (fallacies)? What about rhetoric?

In this example, at this stage, the reasons appear to logically lead to the conclusion, though this may change when the quality of the reasons is examined (see below). At this stage, no fallacies are detected. There is some rhetoric favourable for the dogs, with them labelled as 'street' or 'free-living' rather than 'pest' or 'vermin'. Language has powerful connotations, influencing what is deemed acceptable treatment.

A logical argument can still be of poor quality if the reasons are poor. Reasons are made up of facts, values and assumptions. When evaluating the reasons, firstly, resolve the ambiguity in the words so that everyone is referring to the same thing.

b. Assess the facts.

Facts are supported by evidence. Reading and understanding scientific literature, as well as identifying credible sources of evidence, are essential skills for evaluating facts accurately. Avoid cognitive bias – do not selectively pick evidence that supports your preconceived stance on an issue.

In this example, the World Organisation for Animal Health (WOAH) and the International Companion Animal Management (ICAM) Coalition[58] are credible sources of evidence regarding street dogs. To briefly summarise, killing healthy street dogs by itself is not effective in reducing or preventing rabies cases in humans and dogs in many contexts, as other dogs will move into the area, making use of the surplus food and shelter. Instead, maintaining a stable street dog population often reduces disease transmission by limiting the migration of infected dogs into an area. However, killing healthy dogs can be a useful adjunct to other rabies control measures in certain situations.

The argument is modified according to this assessment of the facts.

Reason 1: 'Killing street dogs does help to get rabies under control when done in conjunction with other actions by citizens, such as rabies vaccination, neutering and more effective garbage control'.
Reason 2: 'We should consider doing things that work'.

....

Conclusion: 'We should consider killing street dogs in conjunction with certain other actions by citizens, such as vaccination, neutering and rubbish control'.

When stakeholders continue to argue against euthanising street dogs, there will be other implicit or unspoken reasons. To avoid confusion and ensure productive discussions, it is essential to identify and understand these. Examples are Reasons 2 and 3.

Reason 1: 'Killing some street dogs in association with other measures, such as rabies vaccination, neutering and rubbish control, can work to reduce the incidence of rabies infection in humans and dogs'.
Reasons 2: 'The other measures can work, even without killing some street dogs, albeit more slowly'.
Reason 3: 'Where it is possible to achieve the same aim without killing dogs, then we should not kill dogs'.

....

Conclusion: 'We should not kill street dogs'.

Note: When evaluating facts (scientific evidence), 'absence of evidence' is not the same as 'evidence of absence'. The 'precautionary principle' often guides animal welfare policy, i.e. preventative measures should be taken in the face of potential harm, even if the scientific evidence is not fully conclusive. Here, a decision is made that gives the animal the benefit of the doubt.

c. Assess the values.

How do the reasons align ethically? Ethical theories are helpful here. Identify which theory best fits and how other theories would modify the proposed action.

Consider Reason 3 from the example, 'Where it is possible to achieve the same aim without killing dogs, then we should not kill dogs'.

This appears to align with the 'ethics of principle', as the street dogs are assigned an inherent value or 'right to life' and an interest in staying alive that should be respected. The 'fairness' of what is being proposed will raise further ideas that might otherwise not be considered. For example, is choosing to kill some healthy animals and not others fair?

'Ethics of consequence' will make us think about the greatest good for the greatest number. Killing some street dogs in the short term may benefit the overall dog population in the longer term, leading us to modify Reason 3 so that the killing of dogs is avoided 'wherever possible'. The phrase 'wherever possible' then needs to be more clearly defined.

'Ethics of purpose' should make us consider the desired qualities of veterinary professionals and what they should be trying to achieve. Examples include making animal welfare our overriding consideration, avoiding killing healthy animals wherever possible, and using our knowledge to protect animals and humans from serious zoonotic diseases.

d. Assess the assumptions.

Assumptions are the 'taken-for-granted notions' of how the world tends to work, for example, how people behave or what tends to happen in certain situations. If people agree on the facts and the values underpinning the proposal but still disagree on the proposed course of action, then they must hold different assumptions.[47] For example, some people may automatically assume that the lives of street dogs are less valuable than those of other dogs because they do not have owners. Some people may assume, without further inquiry, that street dogs prey on rats and that lowering street dog populations will lead to problems with rat infestations.

5. Modify the proposal.

By this stage, participants should have clearly identified what they agree and disagree on, and why. They can modify the proposal based on what has been learned, hopefully making the best possible decision whilst keeping an open mind about later revision.

This example has purposefully been kept very brief for illustration purposes – in reality, there will be much more nuance. It also highlights the dangers of relying mainly on economic arguments or the fact that 'something does not work' to justify actions that protect animals. If the argument falls short, now or in the future, animals will no longer be advocated for. Robust justification properly considers many additional elements.

Fallacies and rhetoric

Identifying 'logical fallacies' in reasoning is crucial, as they can undermine or invalidate an argument. A full discussion is outside the scope of this book, but it is worth consulting the subject area of 'critical thinking' to become better at identifying and quickly responding to them. In brief, fallacies are errors in reasoning (bad arguments) that can, nevertheless, be highly persuasive.[57] A few examples of fallacy types are as follows:

- **'Conflating morality with legality'** is the mistake of assuming that anything legal must be moral and, conversely, that anything illegal must be immoral. The 'there's no law against it, so it's acceptable' argument is extremely common.

- The '**majority belief fallacy**' asserts that something should be done or believed *solely* because it is what the majority of people want or believe.
- The '**appeal to ignorance**' fallacy asserts a claim must be false *solely* due to a lack of proof. However, the absence of evidence does not equal evidence of absence; the topic may simply require further study.
- '**Mistaking correlation for causation**' occurs when a statistical correlation between two things is assumed to mean one thing causes the other, without further justification or investigation to test this.
- The '**ad hominem**' fallacy discredits an opponent of a position (for example, by labelling them as biased) instead of addressing the actual argument. (Conversely, some accept arguments *solely* based on the authority of the speaker rather than evaluating the facts, values and assumptions – see 'Appeal to authority'.)
- '**Appeal to authority**' is when someone's expertise or credentials are used with the assumption that their argument is *automatically* correct.
- The '**straw man**' fallacy is where someone's opinion or position is misrepresented, exaggerated or oversimplified in order to make it easier to attack.
- The '**slippery slope**' fallacy is where an argument is made that a relatively small first step always, and inevitably, leads to a series of worsening events, without demonstrating any logical or realistic connection between these steps.
- A '**false dilemma or false dichotomy**' is where a situation is presented as having only two possible options when, in reality, there are more.

Rhetoric is a persuasive technique that does not rely on logical reasoning but simply uses language to persuade by creating a certain feeling or belief. Language can be powerful, loaded with connotations that influence how we act. Examples include:

- **Emotional appeal**, where emotive language is used to tap into people's emotions, such as their empathy or to stir up feelings of anger and injustice.
- **Logical appeal,** where statistics, numbers, graphs, etc. are used to create meaning that the data, upon proper analysis, do not support.
- **Credibility rhetoric**, where the language used undermines an opponent or enhances the perceived credibility and trustworthiness of oneself or a proponent.

Law

5

Abstract
This chapter intends to provide a basic understanding of the law – its strengths, limitations and peculiarities – laying a foundation for veterinary professionals to delve into more advanced sources of information. The Animal Welfare Act (AWA) 2006 of England and Wales is discussed as an example of a well-developed, national animal welfare law. World Trade Organization (WTO) rules and European Union (EU) laws are used to illustrate how international trade laws can impact animal welfare.

Introduction

At its most basic, law is a system of rules that governs everyday life. Every society in every age has needed a set of rules to live by. These rules usually restrict some of our freedoms but provide us with protections in return, such as protecting us from the harmful behaviours of others. Laws are not the only rules that govern societies. There are also social and moral rules. However, laws often carry much greater force.

The law should not be confused with ethics. Some principles of morality are given the force of the law, and others are not. In democratic societies, laws often reflect a compromise between viewpoints and will not please everyone. Some issues are clear-cut. For example, most societies find murder abhorrent and treat it as a criminal offence. Other issues are more controversial, including how we relate to animals.

Not everything legal is regarded as ethical by everyone, and many things that can be regarded as ethical are not supported by the law. Where laws are created, they generally reflect the minimum standards required as opposed to best practices or the ideal situation.

Law has a cultural dimension. Understanding culture helps us to understand why different laws have developed over time and why different legal traditions have developed in different countries.

DOI: 10.1079/9781786391582.0005

In many countries, there are set processes and procedures by which values become law, and there are defined institutions with the authority to create laws.

The law is only effective if understood, adequately implemented and enforced. Therefore, knowledge of the law is of limited value unless you understand how the legal system works to enforce the law.

Legislation has its shortcomings. This may lead some to dismiss it as ineffectual because it does not consistently deliver the desired outcomes for animals. However, it is crucial to society as, without it, no formal action could be taken against those who mistreat animals. Equally, it must be noted that simply legislating can never be a panacea, particularly if resources for implementation and enforcement are limited.

Veterinary professionals can use the law to better advocate for animals if they understand how the law approaches animal welfare and how to work effectively within existing legal structures, as well as channels for changing laws.

National and international laws

National laws

In basic terms, a country's government creates national laws that apply to all citizens within its borders. There are different types of national laws. For example, the UK has criminal laws and civil laws. Criminal laws concern the relationship between citizens and the state. Their primary purpose is to punish offenders, and there are penalties for those who break the law. State law enforcement agencies, such as the police, are involved. Examples of criminal laws in the UK that protect animal welfare are the Animal Welfare Act (AWA) 2006, which is the leading animal welfare law of England and Wales (there are equivalent laws in Scotland and Northern Ireland), and the Veterinary Surgeons Act 1966 which strictly regulates who can practise veterinary surgery in the UK, thereby protecting owners and their animals from incompetent persons. In contrast, civil law is used by one citizen against another, often to sue them for breaching a duty of care or a contract. Civil law primarily aims to compensate a person for their loss, which is a private matter and generally does not involve state enforcement agencies. An example is a client suing a vet for alleged negligence.

Numerous countries have enacted laws to safeguard the welfare of animals, although the degree of protection they offer differs. A discussion of the animal welfare laws of different countries is outside the scope of this book. Instead, this chapter will delve into the British legal approach to protecting animal welfare, highlighting its strengths, limitations and peculiarities. Some of these are related to the British cultural approach to animals, while others are inherent to what a legal system can achieve. Key features of the Animal Welfare Act 2006,[59] which applies in England and Wales, are discussed as this is a good example of a well-developed, national animal welfare law. (Note: The United Kingdom comprises three legal systems: England and Wales, Scotland, and Northern Ireland. Additionally, animal welfare is a devolved matter, meaning that each of the four nations can pass its own animal welfare laws. Whilst they have a very similar overall general approach to legislation passed to protect animal welfare, there are also important differences.)

The AWA 2006 resulted from a complete legislative overhaul, replacing outdated laws that were no longer fit for purpose. This process took several years and involved thorough consultation and scrutiny. This was a remarkable feat considering that animal welfare competes for attention with many other important national issues for the necessary political will and parliamentary time.

The AWA 2006 can serve as a helpful reference point for developing animal welfare laws in societies with similar cultures. Indeed, animal welfare organisations have sometimes created 'model laws' for protecting animal welfare that are very similar to the AWA 2006. However, whilst animal welfare charities may encourage their adoption by other countries, simply copying and pasting laws from one country will not be effective or appropriate. To ensure success, laws must reflect the unique culture, ethics, social behaviours and institutional structures of the countries where they are being implemented. Model laws can, however, provide different countries with useful discussion materials.

International laws

In basic terms, international laws are created by agreements between sovereign nations, akin to legal contracts between countries. These countries agree to relinquish a portion of their sovereignty to reap the benefits of the agreed-upon rules. Generally, there is no higher authority, such as an international police force, that can force countries to comply. Even so, nations tend to respect these laws due to their legitimacy, respect for reciprocity and a desire to keep a good reputation, all of which attract business. Sometimes, there are mechanisms in place to encourage compliance, such as applying sanctions to countries that violate rules.

International laws are necessary for regulating relations between countries, ensuring economic and legal certainty in the global economy.[60] The dominant form of international law is the 'Treaty'. International law has established intergovernmental organisations with the power to carry out various tasks on behalf of countries. These are forums for debate that often play an important role in formulating international laws and sometimes help monitor compliance. Notable examples are the United Nations (UN) and the World Trade Organization (WTO). The World Organisation for Animal Health (WOAH), previously known as the Office International des Epizooties (OIE), is a key global inter-governmental organisation established in 1924 to combat animal diseases. Since 2002, it has explicitly recognised animal welfare as a key component of animal health and has been developing science-based standards at an international level, covering terrestrial animals and fish used in farming, slaughter, transport and research, as well as stray dog population control methods.[9] Implementation is highly variable amongst the 183 member countries, and citizens of countries with relatively high welfare standards may criticise WOAH's efforts for not going far enough. However, even gaining the commitment and aspirations to address animal welfare among so many countries with very different outlooks and socio-economic statuses is a major achievement. WOAH provides them with credible policy direction. It also establishes animal welfare as a legitimate global issue amongst other powerful organisations, for example, the WTO, which views WOAH as a credible reference organisation.

Many organisations outside of governments also significantly influence international laws despite lacking formal authority. Included are large companies that wield influence due to their economic importance, and non-governmental organisations (NGOs) that may be recognised for their expertise and principles.[60] NGOs are generally non-profit organisations that further some social or humanitarian issues. Examples include Greenpeace or Human Rights Watch. In the UK, animal protection charities are often called NGOs when operating in other countries. Some focus on improving animal welfare, others advocate an end to animal usage ('animal rights' organisations), and some do both, with their approaches often differing according to the animal sector. Other civil society organisations include professional associations, such as national and international veterinary associations. There are also religious groups. When collectively mobilised, these non-governmental organisations can influence elected governments, companies and the general public.

To engage in global trade in today's interconnected world, a country usually has to comply with certain international laws – national governments do not have free rein to unilaterally determine all their national laws.

Currently, there are no international laws with the primary aim of protecting animal welfare, though some do this as a secondary objective. For example, many trade laws developed by the European Union include provisions specifically to protect animal welfare with the primary purpose of ensuring a 'level playing field' for trade among countries, addressing the fact that raising animals to higher welfare standards is often more costly and can put those doing this at a trade disadvantage (see Chapter 3). International laws designed for conservation purposes, such as the 'Convention on International Trade in Endangered Species of Wild Flora and Fauna' (CITES),[61] indirectly provide some welfare protections in areas where welfare and conservation improve in tandem.

The Animal Welfare Act 2006

The Animal Welfare Act (AWA) 2006, which applies in England and Wales, exemplifies a well-developed, national animal welfare law suited to British culture and institutions.

The following aspects of the AWA 2006 are considered:

- Which animals are protected?
- Who is held responsible/who can commit an offence?
- What are the main protections for animals/what are the offences?
- How is the law implemented and enforced?

Multiple factors have determined the answers, including ethics, scientific knowledge, cultural norms, habitual usage of animals, human priorities, politics, pragmatism and resource availability. Animal welfare is a controversial issue, with many stakeholders having different interests.

Animals protected by the AWA 2006

For laws to be workable and interpreted consistently and fairly, clear definitions are needed.

Therefore, the AWA 2006 law clearly defines which animal species are protected and in which contexts. Both animal and human factors informed these decisions.

It is positive that a key consideration is animal sentience, as this forms the basis of welfare concerns. However, the AWA 2006 only covers vertebrate animals despite strong evidence that some invertebrate species are sentient, as reflected in the recent Animal Welfare (Sentience) Act 2022 covering decapod crustaceans and cephalopod molluscs, and the Animals (Scientific Procedures) Act 1986, which covers octopuses. The AWA 2006 explicitly states it can be extended to include invertebrates and animals in their foetal and larval stages if scientific evidence demonstrates their sentience. However, laws usually change slowly. They also often prioritise the interests of those who benefit from using animals.

A key human factor is the level of responsibility society believes different people should have towards different animals in different situations, encompassing the extent to which the animal relies on care from particular people. The law should be reasonable, bearing in mind that failure to comply can lead to severe consequences such as criminal charges.

Under the AWA 2006, only animals that are owned, under a person's control or commonly domesticated in the British Isles are protected. This includes animals with owners and free-living unowned domesticated animals, such as stray dogs and cats. It also includes wild animals under human control, even temporarily, such as those in wildlife sanctuaries, receiving care in veterinary clinics or caught alive in traps. It is understandable that the AWA 2006 does not cover free-living wild animals. However, the interests of these animals could be much better protected by other laws than is currently the case.

The AWA 2006 inevitably reflects British societal attitudes towards animals. For instance, they are considered the legal property of their owners under 'common law', with the AWA 2006 simply placing restrictions on what people can legally do with their 'property'.

Protected animals have no legal right to life. Instead, the focus is on ensuring an acceptable quality of life and a humane death. In the UK, owners are often legally permitted to kill their animals without justification, even young, healthy animals, provided they do this humanely. This contrasts with some European countries, where a clear justification is required before healthy dogs can be killed,[62] or in some states of India that prohibit the killing of any cow, even those moribund.

The law reflects the moral inconsistency of British citizens. For example, attention to welfare is tempered by animals' utility value rather than a morally relevant factor like their capacity to suffer. Using Peter Singer's philosophy, this can be seen as a type of 'speciesism'. Animal species covered by the AWA 2006 are exempt if used in research, hunting and fishing, though other laws do provide some of them with some protection. Some species may not have the level of protection the law's wording appears to provide, as the courts will often interpret the law as the requirement to meet current industry standards. For instance, the AWA 2006 covers both pet dogs and farmed pigs, but the welfare standards considered acceptable for pet dogs are often greater than those for many farmed pigs. There is robust scientific evidence that the fast-growing strains of chickens used in most conventional farms experience significant welfare issues, particularly lameness. This is the industry norm despite the law stating that 'Animals may only be kept for farming purposes if it can reasonably be expected on the basis of their genotype and phenotype that they can be kept without any detrimental effect on their health and welfare'.[63]

Due to our cultural affection for dogs, cats and horses, they have special protections under UK laws designed to protect animals used in research.

Responsibilities

The AWA 2006 applies to everyone in England and Wales, prohibiting certain actions that harm animals. However, those who own or care for animals have extra legal responsibilities towards them. Here, the law covers neglect resulting from failure to act, not just direct actions that harm animals. Owners cannot absolve themselves of responsibility for an animal's welfare, even if they delegate day-to-day care to others.

Main offences under the AWA 2006

Two main offences under the AWA 2006 are causing a protected animal 'unnecessary suffering' (in section 4) and failing a duty of care to meet the needs of an animal (in section 9).

Causing a protected animal 'unnecessary suffering'

This offence addresses some weaknesses and retains some strengths of prior animal protection laws, that is, the Protection of Animals Act 1911 and its judicial interpretations. It is a versatile offence that covers many situations. At its most basic, it involves asking two main questions: Did the animal suffer? Was the suffering necessary? (The word 'cruelty' is no longer used.) However, it can also be confusing. Unlike other criminal laws, 'suffering' is not explicitly defined, except to acknowledge that it includes physical and mental elements. Other criminal laws use the words 'suffering from', with a clearly defined object when defining offences. The lack of clear guidance can lead to conflicting interpretations and compromise the fairness with which the law is applied. Veterinary professionals involved in animal welfare cases should be mindful not to conflate the everyday meaning of the word 'suffering', which implies something extreme and unnecessary in and of itself, with the legally intended meaning. The definition of suffering provided by Donald Broom, a renowned animal welfare scientist, can provide helpful guidance for evaluating whether suffering has occurred: 'One or more bad feelings continuing for more than a short period'.[64]

Additionally, simply causing 'suffering' is not a criminal offence; the suffering must also be deemed 'unnecessary'. The law provides guidance on factors to consider when evaluating the necessity of suffering. Importantly, the AWA 2006 objectively measures an owner's actions against reasonable standards of a competent, caring owner rather than subjectively according to their knowledge or intent. Therefore, people can be held accountable despite claiming ignorance or that they did not intend harm.

Conversely, the word 'cruelty' implies causing intentional harm. The media and others will often label persons convicted of causing unnecessary suffering as 'cruel'. Veterinary professionals can communicate a more balanced perspective where warranted. For instance, some farmers face socioeconomic conditions that harm their mental health, making it difficult for them to address issues. Usually, they do not want their animals to suffer, but are unable to adequately address the issues. In the UK, organisations such as the Farming Community Network[65] can greatly help farmers navigate their situation and are often open to such requests from veterinary professionals, provided the animal owner/keeper agrees to be helped.

Under the AWA 2006, specific actions, such as poisoning and dogfighting, are considered so harmful and unnecessary that they are outright banned.

Failure of duty of care

Individuals who own, care for, or control animals have more legal obligations than other citizens based on the widely held belief that they should actively provide for their animals' welfare needs.

The duties prescribed evolved from the Five Freedoms framework. They have been reworded as 'needs' for a suitable environment, a suitable diet, the ability to exhibit normal behaviour, the option to be housed with or separated from other animals, and protection from pain, suffering and disease (see Chapter 1).

Implementation and enforcement

The effectiveness of a law depends on how it is implemented and enforced, which in turn relies on having sufficient resources. Notably, enforcing the AWA 2006 is not a legal duty but a discretionary power of local and national authorities. Therefore, the extent of enforcement varies considerably between regions.

The AWA 2006 contains provisions that can be used to help citizens meet their legal obligations.

Licensing and registration

Licensing and registration aim to ensure that animal activities are carried out competently and within the law. They currently apply to areas such as racing greyhounds, riding stables, dog breeding establishments, pet shops and zoos. However, expanding these to cover other areas of animal usage, such as farming and general pet ownership, would arguably help raise welfare standards. Local authorities have discretion regarding inspection details, so the effectiveness of licensing and registration may vary between regions.

Education

Education is crucial if citizens are to know their legal obligations and possess the necessary skills to fulfil them. The PDSA Animal Wellbeing (PAW) reports[66] found that while three-quarters of pet owners knew about the AWA 2006, less than half were familiar with it and their legal responsibilities.

In the UK, animal carers are not legally required to undergo training. Exceptions include stock persons, who must be familiar with the official Codes of Practice created under the AWA 2006, and those slaughtering animals on farms, who must demonstrate their competence in official assessments to become certified. Veterinary professionals can further educate. For instance, private veterinary practices could advertise in their clinics that owners have legal responsibilities, not just moral ones. Veterinary professionals working as official regulators could make it easier to find accurate, necessary information online.

The Codes of Practice created under the AWA 2006 can be useful educational tools. They advise beyond legal minimum standards, informing us of good practices. Codes of Practice are not law, so failure to follow them is not an offence. However, failure to follow the codes may be used as evidence where someone is being prosecuted under the AWA 2006. For some species, the Codes of Practice are extensive and relevant, but others require updating, pending allocation of resources. In recent years, there has been a tendency to develop codes through public-private partnerships. This helps ensure the codes stay relevant to their intended industry, but risks welfare requirements being compromised by industry pressures. Veterinary professionals can be useful in negotiations as the animal's advocate, ensuring the content strikes a good balance.

In most countries, established systems are in place to address illegal activities. Individuals and organisations have the authority to investigate suspected law violations, gather evidence, press charges against offenders, conduct hearings to determine whether laws have been broken, and levy penalties when violations are proven. In the UK, veterinary professionals may become involved in welfare investigations and court cases in various ways. They may have welfare concerns about animals presented directly to them or may be approached by RSPCA/SSPCA inspectors or law enforcement agencies for assistance, such as providing emergency treatment to animals under investigation or certifying them for seizure. Veterinary professionals may be called upon to testify in court as a professional or expert witness. Veterinary professionals working for government agencies often have an official regulatory role, including making routine inspections and responding to animal welfare complaints.

Collecting evidence

Evidence is an important factor in deciding whether to prosecute. The strength of evidence determines whether a prosecution is likely to be successful and whether a defendant chooses to plead guilty or opt for a trial. Law enforcement agencies often request that veterinary professionals assist them in situations involving animals. Consequently, veterinary professionals are often important in gathering evidence during a welfare investigation, though their primary obligation remains the welfare of the animal(s).

There are various types of evidence, including oral testimony from witnesses, documentary evidence such as photographs and drawings, material real evidence like blood samples, and circumstantial evidence. A crime scene is where evidence of a crime might be found, such as a location or an animal's body. Someone with statutory powers, such as the police, can declare a crime scene. It is crucial to safeguard the evidence at the crime scene.

Veterinary professionals who collect evidence will benefit from training in this area. They must stay within their professional capabilities and expertise, or they risk damaging evidence. It is important to gather evidence that shows whether the animal was suffering or likely to suffer, as well as the cause and necessity of the suffering. To ensure the quality and integrity of evidence, veterinary professionals must maintain a chain of evidence. This includes maintaining detailed clinical history and examination records, taking samples properly, and accurately photographing and documenting evidence. The source and secure whereabouts of evidence must always be accounted for. Veterinary professionals should always assume that any case they are involved in will be taken to court, even though most cases do not end up there. Some veterinary professionals have trained to become experts in veterinary forensics. This is a broad, multi-disciplinary subject, with animal welfare cases being just one aspect.

The law clearly defines who has legal powers and their limits, safeguarding against the abuse of power. Veterinary professionals should know whether they possess legal powers, what these are and the legal powers of the persons they are assisting. For instance, police officers, local authority inspectors and government-employed veterinary inspectors are usually legally authorised to enter parts of premises to ensure compliance with animal welfare regulations, whereas private veterinary practitioners generally are not. RSPCA inspectors have been very important in ensuring animal welfare investigations and prosecutions in England and Wales, filling a resourcing gap, but contrary to popular belief, they have no more legal rights than an ordinary citizen to enter private premises or seize materials or animals (in contrast to SSPCA inspectors in Scotland). Courts may reject evidence collected by those breaching their powers, deeming it unlawfully or unfairly collected, thus risking the fairness of the trial. Individuals may also be liable for any damage they cause, being trespassers, in contrast to those with legal powers, who are not liable provided they stay within their powers and act in good faith with reasonable grounds for their actions.

Interim strategies to protect animal welfare

Welfare investigations and court trials are often slow, lasting months or even years. During this time, animals remain the property of their owners. However, there are legal provisions to safeguard their welfare in the interim. 'Official inspectors' have the power to take possession of animals that are experiencing distress or suffering and to take steps to quickly alleviate that suffering, including euthanasia based on veterinary advice. They can also remove animals that are kept in conditions likely to cause suffering, rather than waiting until they have suffered. However, enforcement bodies are cautious about taking this action where significant numbers of animals are involved, as there are substantial costs associated with the animals' upkeep.

Prosecution

In the UK, a charge is a formal, but not yet proven, allegation that an offence has been committed. Prosecution is the conduct of legal proceedings to determine if a charge is proven. It generally takes the form of a trial in court, an impartial system for determining if the person charged broke the law. Trials held in the UK are 'adversarial', with the prosecution and defence pitted against each other when presenting their arguments and evidence to the court. The prosecution and defence are usually represented by their own solicitors and barristers, whose job is to portray their client's case in the best possible light. The court decides which facts are proven and how the law applies to these facts. The burden of proof is on the prosecution, with the defendant innocent until proven guilty. The level of proof required to convict someone of a criminal offence is 'beyond reasonable doubt' compared with the lesser standard of 'the balance of probability' in civil proceedings. A trial can be avoided if a defendant pleads guilty at the early stages of the criminal process. In this case, the court can move to sentencing without conducting a trial. A defendant pleading guilty early will usually benefit from a lower penalty (sentence) than if found guilty after a trial.

Different countries have different legal systems and types of trials. For instance, the 'inquisitorial system' is used in European countries such as France and the Netherlands. Here, the courts actively investigate the facts of the case rather than simply acting as an adjudicator of those presented to them by the prosecution and defence.

Witness evidence

Witness testimonies are part of the overall evidence submitted to the court. Witnesses are called to court to give evidence under oath. In the UK, witnesses can be split into two broad categories: witnesses of fact (including professional witnesses) and expert witnesses. Witnesses of fact will explain what they did and why, but are not allowed to give an opinion. Professional witnesses are witnesses of fact, but in addition, can explain what they did and why from the viewpoint of a professional. Inevitably, they will give some opinion on the matters they have been personally involved in, as their opinion is very closely tied to the facts – their opinions and judgement influence what they did or thought, and the reasons why. Expert witnesses, however, are allowed to give their opinion, even on matters with which they had no personal involvement. Their role is to assist the court on matters beyond its expertise. The prosecution or defence may engage them, but unlike solicitors or barristers who represent their clients, they are not there to be partisan and support the client's case. Their overriding duty is to the court, and they should avoid bias due to external pressures, with their view being the same regardless of which side engages them.

The court decides whether someone is a competent expert witness, often considering their education, academic qualifications and practical experience. However, many people outside the profession assume every veterinary surgeon is an expert regarding animal matters. Therefore, veterinary professionals should behave responsibly and not accept the mantle of 'expert' for areas beyond their expertise. In the UK, expert witnesses are not immune from being sued for negligence.

Penalties for convicted persons

If found guilty (convicted) of an offence by the court, the offender will be sentenced by the court. One purpose of conviction is to punish the offender. At the time of writing, maximum sentences for causing unnecessary suffering to an animal include an unlimited fine and/or five years imprisonment (Animal Welfare (Sentencing) Act 2021). The other purpose is to protect animals. The court can deprive an offender of their animals and disqualify (prohibit) them from keeping animals in the future. Any licences or registrations needed for animal activities will be cancelled. Importantly, they can ban offenders from being involved in any arrangements that allow them to control or influence how animals are cared for, such as other people's animals. Courts need to recognise the significance of disqualification and deprivation in safeguarding animals rather than seeing them simply as alternative methods of punishing the offender. Under the AWA 2006, if the court chooses not to use the deprivation and disqualification options, it must say why.

Cross-compliance and Statutory Management Requirements

Prosecutions under animal welfare law are costly and take considerable time, draining limited resources. Therefore, alternative, faster, cheaper strategies to achieve a result for animals are helpful. An example is the 'cross-compliance' system, which addresses breaches of farm animal welfare law without resorting to criminal proceedings. Following Brexit, this system was radically reformed in England, Scotland and Wales.

The cross-compliance scheme requires farmers who receive subsidies under the EU Common Agricultural Policy (CAP) to agree to unannounced inspections. If they are not meeting certain welfare criteria (Statutory Management Requirements), their financial support is reduced as a penalty. Inspectors classify law violations according to the severity of harm and the farmer's intent. The standard of proof – 'on balance of probability' – is lower than in a criminal court.

International laws – WTO rules and EU laws

Summary

WTO rules are key laws governing international trade globally, though no countries trade solely according to WTO rules. There are now many additional trade laws developed among countries, such as free trade agreements (FTAs) or preferential trade agreements (PTAs), that overlay WTO rules.

Global trade arrangements and supporting laws have evolved over time. In disputes between countries, WTO rules have mostly been interpreted in ways that favour free trade over animal welfare. However, in some disputes, animal welfare has received some recognition as a legitimate trade concern.

The European Union (EU) is unusual in creating FTAs that emphasise protecting socially important issues like animal welfare. In contrast, many FTAs, such as the Comprehensive and Progressive Agreement for Trans-Pacific Partnership (CPTPP), pay little regard to animal welfare.

Note: The international rules-based order for world trade, centred around the WTO, has been in place and has been gradually evolving for several decades. However, at the time of writing, some agreed-upon rules are being ignored by some countries that were instrumental in its creation and the future direction of travel is unclear.

World Trade Organisation rules

WTO rules are key laws governing international trade globally that most countries have agreed to. The WTO is a supranational body through which member countries negotiate and make collective decisions about these laws. The primary goal of the WTO rules is to promote free trade by eliminating trade barriers. They aim to reduce or remove 'border measures' such as customs duties and quotas that national governments may impose on goods as they are imported into their country. There are also two main 'non-discrimination' rules. Firstly, the National Treatment provision prohibits countries from having national rules that favour domestic products over imported ones. Secondly, the Most Favoured Nation provision prohibits countries from having national rules that favour imports from one foreign country over similar imports from another foreign country.[67]

The WTO has a process for resolving disputes when one member country accuses another of rule violations. The primary goal of dispute settlement is to determine if the accused country is violating WTO rules, and then for the offending country to modify its behaviour accordingly. There are several stages to a dispute, and few escalate to the final stage, which is retaliation. Whilst a country cannot be forced to comply with WTO rulings, retaliation can be a powerful mechanism. Here, the complaining country is allowed to apply sanctions to the offending country. Usually, sanctions target influential sectors, motivating them to lobby their government for compliance with WTO rules.

WTO rules in relation to animal welfare

The WTO dispute settlement bodies have usually interpreted WTO rules in a way that prioritises free trade over moral issues important to some societies, such as animal welfare. For example, animals raised to very different welfare standards have been considered similar products ('like' products) under WTO rules because they are judged according to their physical features and economic competitiveness in the marketplace, not their production method. This means that an egg from a low-welfare system is considered similar to an egg from a high-welfare system, as there is no discernible physical difference, and they compete in the marketplace. These supranational WTO rules significantly impact the national laws and policies that governments of member countries feel they can pass. It makes it difficult for countries to discriminate against animals and animal products from countries with lower welfare standards. Whilst the WTO rules allow for some exceptions to 'protect public morals',[67] unfortunately, efforts to uphold animal welfare standards have generally been interpreted as disguised trade protectionism, as illustrated by the reports of several WTO dispute settlements.

Example 1

In the late 1980s, the USA passed a law banning the importing of shrimp caught without the use of Turtle Excluder Devices (TEDs). This law was passed because accidental capture in fishing gear, particularly from shrimp trawling, is a major threat to sea turtles, significantly impacting their populations in the Gulf of Mexico. By 1978, all sea turtle species in this region were listed as endangered by National Oceanic and Atmospheric Administration (NOAA) Fisheries. TEDs protect sea turtles from accidental harm during fishing. They are metal grids fitted into trawl nets, allowing small animals like shrimp to pass through while preventing larger animals, such as turtles, from being trapped.[68]

India, Malaysia, Pakistan and Thailand collectively complained about this law as it affected their exports. Despite acknowledging the conservation and welfare role of TEDs, the WTO Dispute Panel deemed the law 'unjustifiable' and 'arbitrary' trade discrimination since shrimp caught with and without TEDS were considered 'like products', i.e. they are materially the same, simply acquired by a different process.[69]

Example 2

The USA passed a law that applied to both the domestic American fishing fleet and countries that catch yellowfin tuna in the Pacific, whereby tuna caught using the 'setting-on dolphins' technique could not be labelled as 'dolphin-safe'. This law was passed to protect dolphins and prevent consumers from being misled about tuna provenance based on the following. When yellowfin tuna are commercially fished in the Eastern Pacific using purse-seine nets, the schools of dolphins that swim above them risk being accidentally captured and harmed. One technique for catching tuna is called 'setting on dolphins'. Here, dolphins are chased down with speedboats, herded into a tight group, and the purse-seine nets are set around them because the tuna stay with the dolphins and are readily caught.[70] Whilst dolphins are subsequently released, some remain trapped in netting and die or are set free whilst injured and debilitated.

In 1991, Mexico complained to the WTO about the US law, as it hindered Mexico's exports. The WTO dispute hearing accepted the main points raised by the USA in support of their law, but still found the law violated WTO rules as it only addressed risks to dolphins from certain fishing techniques, not all fishing techniques, making the law 'discriminatory'.[71]

However, the WTO approach appears to be slowly changing as animal welfare develops into a well-established and respected discipline. For example, the EU now has laws that ban selling cosmetics tested on animals and the importation of fur from countries using leg-hold traps to catch the animals. Initially, these laws were not implemented, despite strong demand from EU citizens, due to fears that they would not comply with WTO rules.

In 2009, the EU banned the importation and sale of seal products within its market. This decision was made in response to widespread moral outrage among European citizens regarding the inhumane killing of seals. Canada and Norway claimed that this EU law, designed to protect the welfare of seals, violated WTO rules. The resulting EU-Seal Regime case (2014) tested the 'public morals' exception,[67] which is one of a list of government policy goals under WTO law (Article XX of GATT) that allow governments to potentially violate the general WTO free-trade principles. Whilst some aspects of the law were found to violate WTO rules, the WTO upheld the principle that certain national laws and policies can be justified under WTO rules where they protect animal welfare, setting an important precedent.

Potential ways of influencing WTO rulings

The WTO favours national laws and policies that are underpinned by international standards, guidelines and recommendations. National veterinary associations can promote awareness of animal welfare among decision-makers by emphasising robust evidence based on animal welfare science. They can emphasise the animal welfare related activities, international guidelines and recommendations of international organisations, such as WOAH, and EU bodies, such as the European Food Safety Authority (EFSA). For example, in the EU-Seal regime dispute, the Panel referred to the EFSA scientific reports as credible sources of information. The Panel also found sufficient evidence that seal welfare is of moral concern to EU citizens due to other animal welfare laws passed by the EU, and EU activities in animal welfare. In addition, the international animal welfare guidelines being developed by WOAH (then, the OIE) support the argument that animal welfare is a legitimate trade concern. Veterinary associations and animal charities can also influence dispute decisions by presenting arguments as *amicus curiae* submissions, which are written submissions by persons or organisations not party to the dispute. The WTO has the discretion to refer to these reports.

Ways of 'bypassing' the WTO free trade rules

WTO rules apply only to the laws and policies of national governments. Private businesses, such as retailers, retain the freedom to choose which products they sell and can use this to improve animal welfare standards. Here, consumer choices play an important role in determining standards.

As already mentioned, free trade agreements (FTAs) or preferential trade agreements (PTAs) are a potential avenue for improving animal welfare, as today no country trades solely according to WTO rules. FTAs and PTAs appear to be inimical to WTO rules by allowing countries that are parties to a trade agreement they develop among themselves to discriminate against trade from countries outside of that agreement. However, from a practical perspective, they have increased in number and gathered momentum enormously in recent decades, as WTO rules alone could not support efficient trade. They are here to stay. Countries have more scope to protect animal welfare standards via an FTA than using the 'public morals' exception under WTO rules, provided they have the political will and negotiating clout. The European Union (EU) is unusual in creating an FTA that protects various aspects of animal welfare.

European Union Law

The EU is an economic, political and legal partnership between 27 European countries at the time of writing. These countries, called Member States, have agreed to relinquish some of their sovereignty in exchange for the benefits of a shared customs union and the single market. Member States agree on shared standards for certain areas, set in law, that create a 'level playing field' for trade among them. These EU laws facilitate trade between Member States while protecting them from competition with countries outside the EU, known as 'third countries'. A customs union eases trade between Member States by removing tariffs between them whilst maintaining a common tariff for goods entering the EU from outside, reducing competition from goods produced to different standards. A 'single market' eases trade between Member States by removing 'non-tariff barriers', that is, the expense and time required to verify that goods meet the required EU standards, such as conducting border inspections or completing certification paperwork. The single market is so much more than a shared set of regulatory standards. It can be viewed as a comprehensive 'regulatory ecosystem', where compliance checks occur at the point of production rather than at national borders, with ongoing monitoring and enforcement overseen by the supranational organisations of the EU. Non-tariff barriers still exist with 'third countries', which are subject to border checks and additional bureaucracy, even if they share exactly the same regulatory standards as within the EU. Note that since Brexit, the UK is a third country.

EU law does not apply to all areas of animal use. The EU's founding laws (Treaties) only authorise passing laws that primarily aim to promote free trade in animals and animal products. Therefore, areas such as private pet ownership are generally not covered. Also, the EU cannot pass laws *solely* to protect animal welfare. However, considerable animal welfare laws have been passed to create a 'level playing field' for trade, as raising animals to higher welfare standards that citizens find acceptable often carries costs. Included are laws created to protect animal welfare on farms, during transport and during slaughter, with the primary aim of promoting the smooth functioning of the single market. There are laws protecting the welfare of animals in zoos in order to support EU environmental policy.[72] There are also laws requiring farmers to comply with certain animal welfare laws in order to receive finance under the Common Agricultural Policy (CAP), which primarily aims to stabilise markets and ensure a fair standard of living for farmers, as well as the availability of produce for consumers. Where EU law applies, member countries cannot ignore it, with EU law taking priority (having 'supremacy') over national laws.

Overall, EU membership enables Member States to maintain laws that support higher animal welfare standards, despite competition from cheaper, lower-welfare products from non-EU countries. The EU also influences animal welfare standards outside its borders due to its large and attractive market, which gives it considerable negotiating leverage. This empowers it to include animal welfare standards in FTAs negotiated with countries and regions outside the EU bloc. There is also the 'Brussels effect',[73] where the rules and regulations created by the EU extend into economic life beyond the EU without being actively imposed as laws, because businesses operating in third countries are prepared to comply with EU standards of production to access the EU market.

The EU, as a bloc, is a member of the WTO, giving EU Member States greater negotiating power and influence on animal welfare matters than if they represented themselves individually, and also in discussions and negotiations with other supranational organisations, such as WOAH. The UK officially left the EU single market and customs union when the transition period ended at the end of 2020. However, many EU laws are still in place as domestic legislation, and the EU will continue to have a significant influence on the UK. In Northern Ireland, the Windsor Framework was developed following Brexit in order to avoid a 'hard border' on the island of Ireland, which means Northern Ireland is still subject to some EU rules that do not affect the rest of the UK.

Development of EU laws

EU supranational organisations fulfil various roles. Key organisations are the European Council, the EU Commission, the Council of the EU and the European Parliament.

The European Council comprises the heads of government of the Member States and sets the political direction of the EU. Most EU legislation begins at the EU Commission, whose members are appointed by agreement between Member States. The Council of the EU comprises one ministerial representative from each Member State and is responsible for turning EU Commission proposals into laws. The European Parliament, comprising MEPs directly elected by member states, does not have the power to create and pass laws. It is simply consulted on laws and has some chance to influence them.

The primary laws of the European Union are called 'Treaties' and establish the broad legal framework required to support EU objectives, including the free movement of goods, capital, labour and services and the establishment of a common agricultural policy. Secondary laws created under the Treaties, largely Regulations and Directives, address the technical details for different areas.

Several general principles underpin EU animal welfare laws. Firstly, lawmakers are legally required to acknowledge that animals are sentient, and 'full regard' must be given to animal welfare when developing policies and legislating.[74] However, the term 'full regard' is still very much open to interpretation.

Secondly, laws are expected to be underpinned by scientific evidence. Sometimes this is legally required. The EU funds numerous collaborative projects focused on animal welfare. The European Food Safety Authority (EFSA) has a team of animal welfare experts who review and summarise the vast amount of animal welfare science for different species in different contexts at the request of the EU Commission. EFSA has produced numerous detailed reports that also provide an overview of different husbandry systems in various Member States, giving the scientific findings context. An example where EFSA scientific reports[75–77] have shaped the content of laws is the legal minimum standards established for farmed pigs. Report findings have led to reductions in the time period that pregnant pigs can be housed in individual stalls or gestation crates, bans on the use of tethers, and requirements for better quality flooring and more living space for gilts.[78]

Thirdly, individual Member States have the right to enact animal welfare laws that exceed or demand stricter standards than those set out by EU legislation. There are no legal barriers to doing so. For instance, the UK implemented national laws in 1999 (i.e. when still an EU member) that banned the use of sow stalls before the EU-wide ban in 2013. However, market pressures make this type of change difficult, as domestic producers must then comply with higher welfare standards and associated costs, which can reduce their competitiveness in the marketplace.

Fourthly, there is also an ethos of consulting EU citizens on what matters to them to shape policies. The Eurobarometer is a polling instrument used by EU institutions, established in 1974, to regularly monitor the opinion of European citizens on various issues. There have been four Special Eurobarometer surveys focusing on the attitudes of European citizens to the welfare and protection of farmed animals, published in 2005, 2006, 2016 and 2023.[79] All 27 EU countries were surveyed as part of the 2023 Special Eurobarometer report. Over 80% of European citizens believe that the welfare of farmed animals should be better protected in their country than it is now[24] and a similar proportion supports limiting the transport time of animals. Around three-quarters of respondents support better protection of the welfare of pet animals in their country, and the majority want farming and breeding practices to meet basic ethical requirements.

Each Member State of the European Union impacts the development of EU policies and laws. By pooling resources and expertise, countries with limited resources and knowledge about animal welfare, or where it is not a top priority, can receive support from other Member States to improve their standards.

Implementation and enforcement of EU laws

Regarding animal welfare, each EU Member State is responsible for implementing and enforcing EU laws within its own borders while being overseen by the EU Commission through the Directorate-General for Health and Food Safety (DG SANTE). While common legal standards are applied to all Member States, inconsistencies exist in how these standards are implemented and enforced, resulting in varying levels of animal welfare across countries. The EU Animal Welfare Strategy for 2012 to 2015 emphasised that a cultural appreciation for animal welfare plays a crucial role in whether the spirit and the provisions of legislation are respected. Some Member States take inadequate measures to inform stakeholders, train inspectors, conduct checks and impose sanctions. The EU has now established Reference Centres for Animal Welfare with the aim of improving the enforcement of animal welfare legislation by providing the necessary support for those conducting official controls related to animal welfare. Amongst other things, they provide scientific and technical expertise, develop methods for assessing welfare, offer training courses for staff from both EU countries and non-EU countries and help disseminate best practices across the EU.

There is a Court of Justice of the European Union (CJEU). Whilst enforcement of EU animal welfare laws is for national authorities, the judges of the CJEU provide guidance on the interpretation of the law and how it should be applied. They provide Member States with interpretations of EU law upon request, known as 'preliminary rulings'.

The Council of Europe

The Council of Europe[80] should not be confused with the Council of the European Union. The Council of Europe is not an EU institution. It is an intergovernmental organisation with 47 member countries, established post-war in 1949 to promote unity among European countries. It produces Conventions to promote human values, including animal welfare. These Conventions have become less relevant over time, being largely replaced by EU laws within the European Union. However, they can still positively impact animal welfare in participating countries, not all of which are EU members.

Channels for effecting change

In most democratic nations, the creation of laws follows a structured process. Initially, laws are proposed as ideas that require political backing to advance. Lawmakers must be persuaded to allocate sufficient time and resources. The facts need to be presented clearly and compellingly to motivate action. Education, campaigning and citizen lobbying play crucial roles.

It is helpful to reflect on how lobbying has achieved legislative changes on behalf of animals.

Cruel Treatment of Cattle Act 1822 (Martin's Act)

Martin's Act of 1822 was the first law in the UK aimed at preventing cruelty to animals. It was considered quite controversial at the time of its enactment on 21 June 1822. The law made it an offence for anyone to 'wantonly and cruelly beat, abuse, or ill-treat' any horse, mare, gelding, mule, ass, ox, cow, heifer, steer, sheep or other cattle. It is noteworthy that this was even before the establishment of the metropolitan police force in 1829. Richard Martin, the founder of the Act, recognised the importance of gaining support from legislators and keeping the public informed and interested. He himself looked for cases and brought them to court in order to garner publicity for his cause and establish precedents in court (Figure 5-1).[81]

Ruth Harrison's book 'Animal Machines'

In 1964, Ruth Harrison's book 'Animal Machines' gained widespread attention in the UK after extracts were published in a national newspaper.[3] This highlighted the welfare of animals living in intensive farming systems and sparked

Figure 5-1. The Trial of Bill Burn Under Martin's Act, Print by Charles Hunt after original by P. Matthews, © Parliamentary Art Collection, WOA 6299.

discussions in parliament. As a result, a wave of farm animal welfare legislation was passed in the UK. It can be argued that it had a significant impact on Europe, ultimately leading to the Council of Europe establishing the European Convention for the Protection of Animals Kept for Farming Purposes in 1976.

In 1964, Mr Hoy, Member of Parliament (MP) for Leith, asked a question in Parliament about the potential to amend animal welfare laws to encompass 'modern methods of livestock rearing'. He drew attention to the fact that '*thousands of people were horrified by the Press picture of the modern method of pig production in this country and that the revelations in Ruth Harrison's book astonished many people who were not aware of what was happening. While people are not opposed to modern methods of cattle and fowl production, they are appalled by the cruelty which is taking place*'.[82]

Lobbying for change

Individual veterinary professionals can advocate for change. For example, in the UK, any individual can formally start a petition as a first step. With the help of mass media and social media, important issues can be brought to the forefront. Petitions with 10,000 signatures receive a government response, and those with 100,000 signatures are eligible for discussion in Parliament.

Individuals can lobby their MP on animal welfare matters. By signing up at 'TheyWorkForYou', it is possible to receive email updates on animal welfare questions asked in the UK Parliament by MPs representing their constituents. Also, MPs can propose new legislation. For example, Chris Loader, MP for West Dorset, put forward the Animal Welfare (Sentencing) Bill, which became law in 2021, extending the maximum penalty under the Animal Welfare Act 2006.

Veterinary associations can be a powerful and respected collective voice. For instance, the British Veterinary Association (BVA) is an important resource for lawmakers seeking guidance on animal welfare issues in the UK. This organisation welcomes contributions from all veterinarians, regardless of membership status, and actively consults them on animal welfare matters and policy positions. In 2016, the BVA prioritised animal welfare by launching its Animal Welfare Strategy, which prioritised six areas, including engaging with legislation. The other five areas were animal welfare assessment, ethics, advocacy, education and international efforts. Progress was evaluated, and areas of action were refreshed in 2025.

Veterinary professionals can campaign for specific animal welfare goals to be part of election manifestos and hold governments to account. For instance, the Labour Party, which won the UK general election in 1997, had made a manifesto commitment to introduce new measures to promote animal welfare and a free vote in parliament on whether hunting with hounds should be made illegal. This culminated in the Animal Welfare Act 2006 and the Hunting Act 2004.

Part 2

Pigs

6

Abstract
The core welfare concepts discussed in Part 1 are applied to farmed pigs. Common welfare hazards are listed. Common welfare outcomes are identified and structured according to the Five Freedoms framework. Practical welfare assessment protocols are referenced. Examples are provided of how veterinary professionals have advocated for pig welfare or could be more involved.

Introduction

Pigs (*Sus scrofa domesticus*) have been popular production animals since they were domesticated from wild boar. They are prolific and efficiently convert a wide range of feed types into meat. Today, vast numbers of pigs are farmed globally. Pig meat comprises around one-third of all meat consumed globally, second only to chicken meat.[83] The top three pork-producing regions are China, the European Union and the USA, with populations of over 450 million, 130 million and 70 million pigs, respectively.

Commercial pig farming systems vary widely across and within countries and can be broadly categorised as indoor, outdoor or combination systems.[9] Indoor systems house pigs in groups or individually, often with very controlled environments; outdoor systems keep pigs in paddocks with access to shelter and shade, grouped by production stage; and combination systems integrate both approaches, such as breeding pigs outdoors, after which piglets are weaned and then reared indoors. There is considerable nuance within each category.

This chapter focuses on the conventional, commercial pig farms that produce much of the pig meat globally. These are mainly indoor systems, though some are combination systems. Husbandry systems have developed to optimise production – animal welfare awareness and knowledge came later. While welfare and production can improve in tandem in some areas, they should not be conflated (see p. 12). Pigs can remain highly productive despite experiencing various welfare problems.

 Animal Welfare: A Concise Guide for Veterinary Professionals (K. Anzuino, J. Anzuino and P. Roger)
DOI: 10.1079/9781786391582.0006

In some countries, pig welfare has been a significant concern for citizens. High-profile campaigns by animal welfare organisations have raised awareness of husbandry practices that severely restrict normal pig behaviour, using imagery of sows in stalls and crates, something that most people can readily understand. Pig welfare has been extensively studied scientifically. Pivotal reviews and reports by relevant EU bodies have informed EU laws, resulting in EU countries having some of the highest standards for pig welfare globally. However, even here, multiple factors make it challenging to improve pig welfare, including generating sufficient, sustainable consumer and government support to make it economically viable for farmers to invest in more welfare-friendly systems.

Husbandry

Figure 6-1 shows a typical production cycle for pigs in conventional, commercial farming systems.

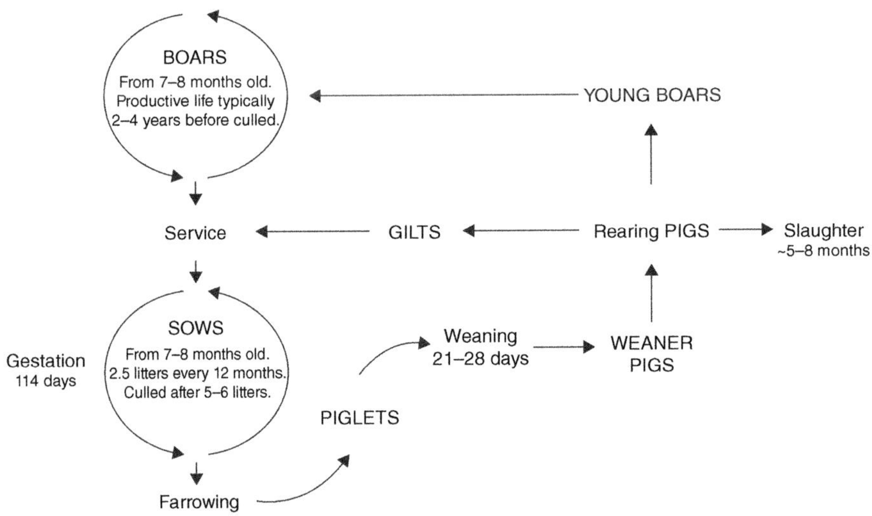

Figure 6-1. Typical production cycle. Definitions – Gilt: a young female who has started her reproductive life but has not yet farrowed. Sow: female pig after her first farrowing. Piglet: from birth to when removed from the mother (weaned). Weaner: a young pig that has been weaned off its mother and is no longer fed milk. Rearing pigs: from 10 weeks of age until killed, or until their reproductive life starts when kept for breeding. Boar: male pig after puberty, retained for breeding.

Stalls ('gestation crates') for sows and gilts

Worldwide, stalls ('gestation crates') (Figure 6-2) are the main housing for pregnant sows and gilts from service to farrowing. They are metal enclosures, around 2 metres long and 0.7 metres wide, with a trough at the front and a rear gate. Flooring is often fully or partially slatted concrete. Bedding is seldom used. EU laws limit their usage in EU countries to a 4-week period after service. Some countries, such as the UK, have banned their usage altogether. Therefore, since 2013, most pregnant sows and gilts in European countries have been housed in groups rather than individually.

Figure 6-2. Stalls ('gestation crates').

Group housing of sows and gilts

The alternative to stalls is group housing of sows (Figure 6-3). Straw bedding is usually provided. Group management varies. Some groups are 'static' (group members stay the same and the whole group moves to farrowing accommodation together). Others are 'dynamic' (newly served sows enter, and those due to farrow leave). Feeding practices vary, for example depositing feed for the group on solid floor areas or providing shared troughs along the side of the pen. In some husbandry systems, sows are individually fed using 'electronic sow feeders'.

Outdoor bred

Conventional pig farming in the UK is unusual in that many breeding sows (approximately 40%) are housed outdoors (Figure 6-4), rather than reserving this practice mainly for organic and niche systems. Outdoor paddocks typically have electric fences and provide wooden or metal shelters. Group sizes and space vary according to factors like climate and soil type. Sows are often fed cereal-based pellets scattered on the ground, with water supplied automatically in troughs. They are typically moved to individual farrowing paddocks about a week before giving birth and remain there until their piglets are weaned. Most piglets are housed indoors after weaning until they are killed or enter the breeding herd.

Figure 6-3. Indoor group-housing for pregnant sows.

Figure 6-4. Sow and piglets kept outdoors.

Farrowing crates

While the use of gestation crates varies by country, many indoor farming systems do use farrowing crates (Figure 6-5), designed to prevent sows from crushing their piglets when confined in small spaces. Sows are moved into crates around one week before they give birth and remain there until their piglets are weaned. Crates usually comprise metal rails, around 2 metres long and 0.45–0.65 metres

Figure 6-5. Sow with piglets in a farrowing crate.

wide, with a feed trough at the front, and an unobstructed area behind the sow for assisted farrowing. As for stalls, the floor is usually partially or fully slatted, allowing excreta to fall into slurry pits below. Water is provided ad lib from nipple drinkers. Piglets have an adjacent, solid floor space that the sow cannot lie on.

Weaner and rearing pig accommodation

Weaners and rearing pigs are usually housed indoors in mixed-sex, single-age groups of between 10 and 40 animals (Figure 6-6). Flooring may be fully or partially slatted, allowing excreta to fall into a slurry pit below. Common flooring materials include plastic, metal and concrete. Feed is typically provided ad libitum, using various feed dispenser/trough designs. Some pens include hanging toys and other enrichment options, with basic requirements mandated by EU and UK law.

Breeding boars

Mature breeding boars are usually housed in individual pens with partially slatted floors and little enrichment. In EU and UK law, it is a legal requirement that they must be able to hear, see and smell other pigs. Boars are also at risk of multiple welfare harms, which are important to address, and these have been reviewed by the EFSA Panel on Health and Welfare.[75]

Common welfare hazards

- Feeding regime.
- Level of confinement and space allowances.

Figure 6-6. Weaners and rearing pigs.

- Flooring type.
- Thermal environment and air quality.
- Level of environmental enrichment.
- Genetic selection for production traits.
- Age and management of weaning.
- Mutilations.
- Social grouping and mixing unfamiliar pigs.
- Handling quality and stockperson competence.

Welfare hazards are the resources (inputs) that can lead to poor welfare outcomes. The likelihood of poor welfare outcomes and their severity (i.e. the risks) can vary. Severity encompasses both duration and intensity. Common welfare outcomes are identified below, structured by the Five Freedoms framework.

Pig welfare is determined by how well pigs can adapt, both physically and mentally, to their environment (see p. 8). Relevant descriptions of 'free-living' pigs are provided throughout to highlight the adjustments they must make while in human care. Here, 'free-living' is defined as pigs living freely in an environment with the resources for which they have evolved and making their own choices. However, such comparisons provide only part of the welfare picture; welfare science (see Chapter 2) is essential for a proper understanding of the welfare impact of different husbandry systems.

Key determinants of welfare are the knowledge, skill and motivations of stockpersons, as well as whether they have sufficient resources (see pp. 24–25). Where these factors are optimal, welfare will be as good as it can be within a given husbandry system. However, there are often persistent welfare issues that are challenging to resolve without altering the husbandry system itself.

Common welfare outcomes

F1. Freedom from hunger and thirst

Farmed pigs are fed primarily to meet production targets, and feeding regimes do not always meet their welfare needs. As a result, some categories of farmed pigs risk experiencing hunger. Examples are:

Restricted quantities of low fibre feed

Sows and gilts that are fed restricted quantities of high-energy, low-fibre, grain-based feed are at particular risk of chronic hunger. They do not feel full after consuming such small quantities of quickly digested feed, despite having adequate body condition scores and their nutritional needs for production being met. In contrast, the diet of 'free-living' pigs is voluminous and varied (see F5).

Early weaning

Farmed piglets are often weaned very early, typically at 3–4 weeks in Europe and even sooner in the USA, to maximise the number of litters a sow can produce annually. They are not biologically ready to transition from milk to a solid cereal-based diet, plus the transition is sudden, so they may be hungry and thirsty around the weaning period, and also risk other issues (see F3, F4 and F5).[75] In contrast, natural weaning in pigs occurs around 3–4 months of age, and they do not eat significant amounts of solid food until they are 4–6 weeks old.

Competition among pigs

Farmed pigs need to be carefully managed so that all animals can access the quantities and quality of food they require. Pigs risk hunger and thirst if their access to feed and water is limited by competition with other pigs or by painful conditions, such as lameness, that affect their affect their mobility.

Group housing can be very beneficial for farmed pigs, as 'free-living' pigs are social animals, living in stable groups, typically of up to 30 individuals. A social hierarchy forms within the group, maintained by a mixture of friendly and unfriendly interactions. This hierarchy can promote maximum cohesion of group members with minimal aggression. Group sizes are determined by the terrain and resources available. In contrast, farmed pigs are confined to specific areas and rely on what their human carers provide. Pigs need to be carefully matched with resources to avoid changing the social dynamics of the herd towards unfriendly encounters and aggression. Subordinate pigs can be hungry and thirsty if they must compete with more dominant pigs for food and water, and this can escalate to physical fights and injuries. Mitigations include providing ample space for all pigs, sufficient quantity and quality of feed, and using pen layouts and feed trough designs that help prevent pigs from being displaced whilst eating, such as shoulder-length or body-length partitions between feeding spaces for group-housed sows.[75]

Farmed sows have been selectively bred to produce very large litter sizes of up to 18 piglets, whereas five or six piglets are more typical for 'free-living' pigs. Weaker, subordinate piglets can struggle to access sufficient milk, as the litter size often exceeds the number of functional teats the sow has, and may be hungry as a result.[75] Careful attention to cross-fostering piglets can help mitigate.

F2. Freedom from discomfort

'Free-living' pigs inhabit a diverse range of habitats, often woodland or alluvial terrains. Common features are reliable forage sources, water, shade and escape cover. They have a choice of lying surfaces, shelter and shade and options to perform basic body care behaviours that make them comfortable. They groom themselves by wallowing and rubbing their bodies against posts and trees to remove excess dirt, organic debris and insects such as lice and ticks. When adult pigs feel hot, they will wallow in mud (Figure 2-2, p.15) or lie in cold, wet places and seek shade. Here, they wet their bodies and cool down as moisture evaporates from their skin. They lack sweat glands, so they cannot cool themselves by sweating.

Facilities provided for farmed pigs are very different. They usually have far fewer choices. However, certain choices and resources are fundamental for their comfort levels.

Thermal comfort and air quality

Well-managed indoor housing can fully protect pigs from adverse weather conditions and provide a comfortable microclimate. However, maintaining a comfortable environment for large numbers of pigs housed indoors at high stocking densities can be challenging. Ventilation and climate control systems, whether passive or automated, must function effectively in all conditions. The temperature and air quality experienced by pigs depend on several interrelated factors – such as ambient temperature, humidity, ventilation, floor type, hygiene and slurry management – and even small deviations in these parameters can be detrimental.

Pigs housed indoors generally risk becoming uncomfortably hot, rather than too cold, especially lactating sows confined in crates. Pigs have an excellent sense of smell, for example, 'free-living' pigs can detect food buried well below ground. A deterioration in air quality, such as raised ammonia levels, can be aversive.

Breeding sows and piglets kept outdoors have fresh air but are more exposed to inclement weather. In cold weather, they risk cold stress, particularly the piglets. Farms should only be sited in areas with suitable climates and soil types that are less prone to becoming muddy. Mitigations include providing well-insulated housing and deep, dry straw bedding.

Opportunity to rest

Pigs need comfortable rest. They prefer to rest in groups, often maintaining body contact, provided they are not too hot. 'Free-living' pigs protect themselves in depressions in the ground, which they sometimes line with plant material. They choose to have separate dunging and lying areas, often several metres apart.

In farmed pigs, interrelated factors such as space allowance, flooring type, temperature and the behaviour of their pen mates determine comfort levels. Some pig categories are at particular risk of discomfort. Risks vary according to management. For example, disturbed rest can be a particular problem in weaners and rearing pigs in crowded pens.

Farmed pigs may lack sufficient space to adopt their preferred postures, particularly given their motivation to synchronise their behaviours. Space requirements tend to increase with increasing ambient temperature. When hot, pigs choose to lie apart on their sides. If too little space, some may be pushed to rest in the dunging areas, something they prefer to avoid. When overly hot, they may even seek to lie in the damper, cooler dunging areas if they lack better options. Lying in the dunging area poses health risks, as well as being aversive.

Other pig categories find it very difficult to become comfortable due to severe confinement. Sows and gilts housed individually in crates or stalls can only stand up and lie down, not turn around, walk or adopt their preferred resting postures.

The hard, unbedded floors widely used for some categories of farmed pigs, whether solid or slatted, are very different to the lying surfaces 'free-living' pigs often choose. The degree of discomfort will vary with factors such as the material, texture and cleanliness. Slatted floors are generally less comfortable than solid floors, but they are widely used because they are a cost-effective way to manage pen hygiene, allowing excreta to fall into the slurry pit below.

Pigs that express abnormal behaviours often disturb other pigs' rest. Examples include 'belly nosing' in weaners and 'tail biting' in rearing pigs (see F5).

F3. Freedom from pain, injury and disease

Disease and injury are often associated with pain and other unpleasant feelings. Some common injuries and diseases of farmed pigs are outlined below to explore their welfare significance; further details are available in numerous excellent clinical texts.

Suboptimal husbandry can predispose to non-infectious disease and accidental injury, which in turn predispose to infectious diseases. Risks vary according to the quality of stockmanship and management. Even where stockmanship is optimal, pigs can have difficulty adapting to certain features of the farming system.

Pain, injury and disease can result from the other Freedoms not being met. For example, if the previously mentioned factors associated with discomfort (see F2) are allowed to escalate, they can lead to disease. Gases such as ammonia can reach levels that irritate the eyes and the respiratory system. Uncomfortably hot pigs can become heat-stressed. Piglets that are too cold can develop hypothermia.

Some routine feeding practices predispose to disease. Feed-restricted sows and gilts (see F1) are at particular risk of developing digestive problems, including painful gastric ulcers. Early-weaned piglets commonly experience digestive upsets due to the sudden dietary change at a time when they have an immature digestive and immune system. Additional stressors include a sudden change in housing and mixing with piglets from other litters. These factors predispose to gastrointestinal diseases. Generally, respiratory and gastroenteric diseases are common in farmed pigs.

Injury from the physical environment

Inanimate physical features of the farm environment can injure pigs. Skin wounds and other soft-tissue injuries are common across all pig categories. For instance, sows and gilts confined in stalls or crates commonly bump into the surrounding metal rails. They can develop pressure sores and bursae from prolonged lying on hard flooring. Painful foot and limb lesions are also common in all categories of farmed pigs, and they contribute to lameness. Causal factors include floor types and quality, such as slats that are too widely spaced, grid apertures that are too narrow, rough or slippery. The hooves of pigs are adapted to move over very different terrain.

Genetic selection that strongly emphasises production traits over those important for health and welfare can increase risks of poor welfare outcomes. For example, pigs genetically selected for lean muscle can be more prone to bone and cartilage problems that predispose them to lameness. Selecting for overly large litter sizes is associated with a general increase in piglet morbidity and mortality, including instances in which piglets are crushed by the mother when she lies down (see F5).

Injury from social interactions or inactivity

Pigs may be injured by interactions with other pigs. Pigs may injure others when performing abnormal behaviours. Weaners subjected to 'belly nosing' (see F5) can develop flank wounds. Outbreaks of 'tail biting' in rearing pigs (see F5), where pigs bite other pigs' tails, cause tail wounds that range from minor scratches to serious wounds in the pigs that are bitten.

Management that upsets social relations increases stress levels in pigs (see F4), and this can be severe enough to lead to physical fights and associated injuries. In group-housed sows, excessive competition for feed can escalate into fights that inflict physical injuries. Mixing unfamiliar pigs is particularly risky and must be well-managed to minimise fighting whilst the new hierarchy is being established. Farmed pigs are often mixed routinely. For example, newly weaned piglets from different litters are often mixed to make the most efficient use of space in the weaner accommodation. Sow groupings change when early pregnant individuals enter and those close to farrowing leave. Mitigations include ensuring ample space and resources, a well-designed home pen layout that allows threatened pigs to retreat, and careful observation, with pigs that are not coping promptly removed. Also, the reproductive stage could be considered; for example, breeding females tend to be more aggressive during oestrus.

Pigs can develop disease and injury due to inactivity. For example, sows and gilts confined in stalls and crates can only stand up or lie down, whereas 'free-living' pigs spend much of their daylight time exercising whilst foraging and exploring. Pigs confined for extended periods can develop weaker legs due to reduced bone strength and muscle mass. Claw overgrowth also occurs as their hooves are not adequately worn.

Infectious diseases

The cleanliness of the farm environment influences whether pigs with the above-mentioned welfare outcomes develop secondary infections. Effective biosecurity is crucial, as it greatly reduces the risk of pathogens entering and spreading within farms. The risk of infectious disease spread is greater where large numbers of pigs are housed at high stocking densities. Key considerations include vectors such as other pigs, workers, visitors, equipment and wild animals. Best practices involve proper quarantine for new arrivals, 'all in, all out' systems, and thorough cleaning and disinfection. Separating pigs of different age groups helps prevent disease transmission from older to younger, more immunologically naïve pigs.

Risks differ by farming system. Outdoor breeding herds are generally more exposed to wild animals and nearby pig herds. They are generally more susceptible to endoparasites than indoor-housed herds. However, there are welfare trade-offs. For instance, well-managed outdoor paddocks allow pigs greater behavioural expression (see F5).

Hospital pens

Injured or sick pigs placed in hospital pens must receive appropriate nursing care and treatment, including prompt humane killing if required to protect them from pain, fear and other unpleasant feelings associated with injury and disease.

Mutilations

Mutilations are defined as procedures that interfere with the sensitive tissues or bone structure, performed for purposes other than medical treatment (see p. 20). Mutilations performed in pigs include tooth reduction, tail docking and castration. Injury, pain and considerable fear/distress occur due to restraint, the procedure itself and the subsequent healing period. Mutilations should not be routine; instead, they should be based on a proper risk assessment for each litter. When performed, operators should be fully trained and competent.

Tooth reduction

Tooth reduction involves shortening the canine teeth of neonatal piglets. This is done to reduce injuries among piglets as they establish their 'teat order', which is where newborn piglets compete to secure priority access to the sow's teats which produce the most milk. It is normal for piglets to inflict superficial wounds on each other's faces at this time. However, wounds can become problematic in farmed piglets because the overly large litter sizes can create greater competition, and wounds are more likely to become infected in the farm environment. During tooth reduction, piglets experience handling stress and pain. Usually, no pain relief is used. There are two main methods: tooth clipping and tooth grinding. Tooth clipping is inherently harmful as it consistently splinters the teeth and damages the gums, even when done according to best practice. Tooth grinding, which blunts the sharp tips of the teeth, is less damaging and the preferred option.

Tail docking

Tail docking was developed to reduce injuries caused by 'tail biting' (see F5). Clippers, a hot cauterising iron, or a scalpel are used to shorten the tails of piglets in the first days of their life. The procedure causes acute pain and distress, and usually, no pain relief is used. Chronic pain, for example, from neuromas at the healing tail tip, has not been quantified. The procedure can be refined – for example, by maintaining hygiene and using cautery rather than a cold method – but remains very painful nonetheless. Practical, effective pain relief protocols have yet to be developed. Tail docking should be based on a proper risk assessment, not routine. Instead, the focus should be on good husbandry practices, particularly effective enrichment (see F5), as thwarted foraging and exploratory behaviours are a key reason why weaners and rearing pigs bite each other's tails. This abnormal behaviour is a welfare concern for both the pigs that bite tails and the pigs that are bitten.

Castration

Many male piglets are castrated within the first week of life to prevent unwanted, injurious, aggressive and sexual behaviours as they near sexual maturity at 5–7 months of age, and to avoid 'boar taint', an undesirable odour in the meat. Castration is often done surgically. The use of pain relief (anaesthesia and analgesia) varies by country. Challenges to provision include financial costs, increased workload, personnel competence and access to drugs. More effective, practical pain-relief protocols need to be developed. Keeping male pigs entire, common in the UK, avoids surgery. This is feasible where market requirements allow pigs to be killed before unwanted behaviours and boar taint develop. However, some markets prefer heavier carcasses, necessitating older killing ages. Immunocastration, which requires 2–3 vaccine doses, may be a practical alternative to surgery in some contexts.

Other mutilations

Other mutilations performed on pigs of various ages are ear notching, ear tagging, tattooing and slap marking for individual or group identification. Outdoor sows may be fitted with nose rings to prevent their rooting behaviour and associated pasture damage. Metal rings are inserted through the upper rims of the snout or the nasal septum. This procedure causes pain at the time of insertion and stress from handling. Additionally, sows experience frustration afterwards because their natural rooting behaviours are restricted (see F5).

F4. Freedom from fear and distress

Fear and distress often accompany pain, injury and disease. They also occur independently.

Interactions with other pigs

Pigs are social animals, so well-managed social groupings that promote friendly social interactions and minimise unfriendly ones greatly benefit their welfare. Situations that upset this balance can cause anxiety, fear and distress, even in the absence of any physical injuries.

Examples include excessive competition among group-housed pigs and situations in which unfamiliar pigs are mixed (see F1, F2 and F3). Additionally, the very large group sizes sometimes used are thought to cause considerable social stress, as there is a limit to the number of other animals pigs can remember.

Social stress does not just occur in group-housed pigs. Sows and gilts that are housed individually in stalls or crates also experience social stress, as they are unable to move away from the pigs confined immediately around them.

Sows in farrowing crates can also experience stress due to limited social interactions with their piglets and other pigs. The sudden separation of sows and piglets during weaning can cause stress for both.

Pigs only choose to be solitary at very specific times, such as sows when giving birth. Isolation can cause significant fear and distress. Injured or sick pigs placed in hospital pens should be able to at least hear and smell other pigs, even if they are not in physical contact.

Human–animal interactions

Pigs are prey species that naturally view humans as potential predators, although they can learn otherwise. Calm, patient and skilled stockmanship is crucial for welfare. However, it can be challenging to sustain, and pigs risk developing a chronic fear of humans.

Each stockperson manages very large numbers of pigs, limiting the time spent with each individual. For example, whilst feeding automation offers certain production and welfare benefits, pigs then have limited opportunities to make direct, positive associations between people and food provision.

Poor-quality interactions are common. Bad habits include slapping pigs and shouting. Even pigs handled using best practices can become fearful if the associated husbandry procedure is aversive.

F5. Freedom to express normal behaviour

Highly motivated behaviours, essential for good welfare, include exploring and foraging, as well as nest building and other maternal behaviours. Thwarting these behaviours leads to frustration, boredom and other unpleasant mental states.

Exploring and foraging

Domesticated pigs share many of the behavioural traits of their wild ancestors. They have evolved for habitats that are varied and complex. Here, they are active for much of daylight time, travelling considerable distances while exploring and foraging. They are omnivores, feeding opportunistically, and can digest plant, fungal, algal and animal matter, as well as insects and earthworms. Year-round, the diet of wild pigs is predominantly plant material (60–100%), though, unlike ruminants, they can only deal with very easily digestible forages. Rooting is a priority behaviour for pigs, and most of what they consume will be at or near ground level. When exploring, they use their sensitive snout to 'feel' objects and will often investigate unfamiliar objects by chewing them.

Some farmed pigs, like sows in well-managed straw-bedded yards and outdoor paddocks, have sufficient resources to perform some foraging and exploration. Other pigs, mainly sows and gilts, housed in stalls or crates, have all their movements severely restricted. Groups of weaners and rearing pigs have more space to move, but are often in barren environments. Many farmed pigs are fed low-fibre, uniform, limited-volume feeds, providing little for them to investigate. While outdoor sows inhabit a more complex environment and have the freedom to move, fitting them with nose rings prevents them from engaging fully with it, as it makes it painful and uncomfortable for them to perform their highly motivated rooting behaviour.

Enrichment

Enrichment should be seen as an essential for farmed pig welfare, not a luxury. To maximise value for the pigs, enrichments should be chewable, easily manipulated, investigable and ideally edible. Loose organic substrates, such as straw, hay and silage, are most effective, followed by sawdust and wood. Toys made of plastic or rubber, metal chains and other non-destructible materials do not satisfy the pigs' needs to chew and root, and they may quickly lose interest. Current obstacles to provision include additional labour requirements, financial costs, biosecurity concerns and slurry disposal systems that cannot handle enrichment substrates.

Nest building and maternal behaviours

Studies show that peri-parturient sows are highly motivated to build nests, even when provided with a ready-made one. 'Free-living' sows show complex nesting behaviours. They separate from the group 1–3 days before giving birth and excavate a depression in the earth, lining it with plant bedding material gathered from the surrounding area. They give birth in this nest and stay within it, or adjacent, for the next 1–2 weeks. During this time, sows leave their piglets only occasionally, keeping them warm through close physical contact and protecting them from predators. They avoid crushing piglets when they lie down by alerting them with an elaborate behavioural ritual. Additionally, the nest provides cushioning. 'Free-living'/wild sows have much smaller litter sizes than farmed sows, and their piglets tend to be stronger. When the piglets are 1–2 weeks old, the sow gradually leads them away from the nest to forage. She continues to lead and protect them, even after weaning.

Farrowing crates

Farrowing crates cause serious welfare issues for farmed sows and gilts, such as thwarted nest building and limited social interactions with their piglets and other pigs. However, they remain a cost-effective way of managing sows and piglets from birth to weaning. These welfare issues can only be resolved by phasing out the use of crates. Many European countries are leading the way, although the use of farrowing crates is still widely accepted worldwide.

Successful implementation of more welfare-friendly 'free farrowing' systems requires considerable time, financial investment and education. For example, stockpersons need support to develop new skills. Breeds/lines of sows that produce smaller litter sizes, more robust piglets and stronger maternal instincts are required.

Abnormal behaviours

Abnormal behaviours are generally a sign of poor welfare – that the animals' environment is deficient in something they need (see Chapter 2). They can develop when highly motivated behaviours are thwarted. For example, thwarted foraging and exploratory behaviours are a key reason why weaners and rearing pigs bite each other's tails (see F3). Pigs in stalls and crates frequently bite the bars and may develop oral stereotypies. 'Belly nosing', where weaners massage each other's stomachs with their snouts, is thought to be a redirected suckling behaviour in piglets weaned before they are physiologically and behaviourally ready.

Welfare assessment protocols

Protocols comprising mainly animal-based measures are relatively well developed, principally the Welfare Quality® protocols[8] for sows and piglets, and for growing and finishing pigs, and the AssureWel protocols[23] for dry sows and for finishing pigs (see Chapter 3).

Advocacy examples

- In 2021, the Federation of Veterinarians of Europe (FVE) published its position paper, 'Moving towards more welfare-friendly farrowing systems'.[84] To support such changes, websites such as Free Farrowing,[85] provide information collated by scientists that pig farmers and vets can engage with to learn more about different free farrowing and lactation systems, including the economics and grant opportunities.
- The FareWellDock project[86] is an EU initiative investigating and educating on the harms of routine tail docking as a preventative measure in tail biting and how the need for tail docking may be removed. It provides multiple fact sheets in several languages.
- The EU campaign 'Alternatives to Pig Castration', launched in 2010, encouraged stakeholders – farmers, meat processors, and retailers – to transition away from the surgical castration of pigs, providing helpful, practical resources such as videos, fact sheets and reports, and encouraging appropriate anaesthesia and analgesia where the procedure is still performed (periodically updated, available at https://food.ec.europa.eu).
- In 2018, the EU Reference Centre for Animal Welfare focused on pigs (EURCAW-Pigs)[87] was set up because EU institutions recognised that effective implementation and enforcement of laws require official inspectors to be properly trained and have ready access to user-friendly information. Their website posts multiple useful resources.
- The EU continues to strengthen the content of EU pig welfare law[78] by ongoing evaluation and updates in line with credible scientific reports on various pig welfare topics. These reports were initially produced by the Scientific Veterinary Committee in 1997 and later by the EFSA Panel on Health and Welfare in 2004, 2005, 2007, 2012 and 2014, with the latest report published in 2022.[75]

7 Chickens

Abstract

The core welfare concepts discussed in Part 1 are applied to farmed chickens. Common welfare hazards are listed. Common welfare outcomes are identified and structured according to the Five Freedoms framework. Practical welfare assessment protocols are referenced. Examples are provided of how veterinary professionals have advocated for chicken welfare or could be more involved.

Introduction

Domestic chickens (*Gallus gallus*) are the most commonly farmed birds worldwide. Chicken is the most widely consumed meat globally, accounting for approximately 40% of total global meat consumption. Over 85 million tonnes of eggs are produced by laying hens,[88] which equates to over 1.5 trillion eggs.

Chickens are well-suited to farming because they are small and fecund, with short reproductive cycles and rapid maturation. Initial commercial production involved small flocks of chickens that had access to outdoor areas, and the same strains were used for meat and egg production. These dual-purpose birds were useful for converting low-quality foods into high-quality protein for humans, but they grew slowly and had poor feed conversion. From the 1950s onwards, production intensified in many countries. Chickens were moved indoors to improve production efficiency, and laying hens were confined in cages, stacked to better use the vertical space in buildings. There are now separate large-scale, commercial egg and meat (broiler) industries that use very different strains of birds. Research developments in husbandry and disease control significantly advanced production. Research interest in welfare developed later, prompted mainly by citizens' concerns over the severe behavioural restriction of caged laying hens. Many aspects of chicken welfare have now been extensively researched. Priority welfare issues vary considerably when comparing laying hens in caged and non-caged systems and when comparing laying hens to meat chickens.

DOI: 10.1079/9781786391582.0007

Husbandry for laying hens

Fertile 'hatching eggs' are produced by breeding birds in the 'parent flock'. They are removed when laid and transported to a specialised hatchery for artificial incubation. Chicks hatch after 21 days. About 50% of newly hatched chicks are male and often killed shortly after hatching in many countries. However, this practice is under review in some countries (e.g. the UK) and is illegal in others (e.g. Germany) due to ethical concerns. Instead, attention is shifting to sexing eggs before hatching and destroying those that contain males.

Female chicks are vaccinated, 'beak trimmed' and transported to rearing accommodation 24–72 hours after hatching, and referred to as 'day-old' chicks during this time period. Immature hens, called 'pullets', stay in rearing accommodation until about 15–16 weeks old. They are then moved to laying accommodation, reaching sexual maturity or 'point of lay' at around 20 weeks. Once they start laying eggs, they are referred to as laying hens and produce 'table eggs', mainly for human consumption.

Hens typically lay eggs until 72–100 weeks old. They lay about one egg per day in any individual clutch pattern, with production targets of around 320 eggs per hen over 72 weeks and up to 500 over 100 weeks. Productivity then dips, known as 'end of lay', and they are transported for slaughter.

Farming systems may be entirely indoors or provide some outdoor access. Cage systems are entirely indoors. They may be conventional (battery) cages (Figure 7-1), now illegal in some countries, including the UK and EU countries, but widely used globally, or they may be furnished/colony cages, otherwise

Figure 7-1. Typical view inside a house containing battery cages.

known as 'enriched' cages. Many non-cage systems are entirely indoors and are referred to as barn systems. They can be a single-tier/flat deck design (Figure 7-2) or a multi-tier/aviary design (Figure 7-3). Some systems provide birds with outdoor access, such as free-range and organic, typically comprising barn system housing with access to an outdoor range (Figure 7-4).

Figure 7-2. Indoor barn system, single tier.

Figure 7-3. Aviary (multi-tier) system.

Figure 7-4. Free-range barn pophole (left). Typical range used for free-range chickens (right).

The proportion of birds in each system varies by country. Recent figures for UK packing stations show that 23% of eggs come from enriched cages, 13% from barns, 60% are free range and 4% are organic.[89]

Conventional battery cages

Battery cages are small, barren enclosures typically housing 4–6 birds, with space allowances of around 450–550 cm^2 per bird. Wire mesh floors separate chickens from their droppings to maintain hygiene.

Enriched (furnished or colony) cages

In enriched cages, mesh floors continue to separate birds from their droppings, but there is more opportunity for behavioural expression than in conventional cages. The main difference between 'furnished' and 'colony cages' is capacity, with furnished cages holding about five birds and colony cages typically holding 20–80 birds. Each bird typically has around 750 cm^2 of floor area, including some solid flooring for scratching. 'Cage furniture' allows some perching and nesting behaviours.

Indoor barn systems

Key features of indoor barn systems are very large group sizes, littered solid floor areas, slatted floor areas with manure belts underneath, perches, artificial lighting, curtained nest boxes, automated drinkers and an automated feed line. Single-age female birds are grouped together, and large farms may have multiple age groups, with each age in a separate house. Aviary systems offer similar facilities but have multiple tiers, using the vertical space within the building.

Free range

Barn and/or aviary systems may be combined with an outdoor range, creating a free-range system.

Key additional features of a free-range system include popholes along the side of the house, allowing access to the range during the day, and fences to contain birds and protect them from ground predators. Desirable features include cover in the form of trees, shrubs or artificial structures, which protect chickens from both real and perceived aerial predation.

Husbandry for meat chickens

A small number of specialist integrated companies dominate most aspects of broiler production. Parent birds (broiler breeders) selected for desirable production traits are bred to produce fertile hatching eggs, which are removed shortly after they are laid and transported to specialised hatcheries for artificial incubation. Eggs hatch after around 21 days. Typically, male and female chicks are transported to broiler-rearing houses when they are 'day old' (some farms are considering transporting eggs to hatch on the site where they will be raised). Once on the farm, the birds are generally referred to as 'broilers' or 'meat chickens'. They grow rapidly on a sequence of starter, grower and finisher diets, and are typically removed for slaughter at 5–6 weeks of age at 2.2–2.9 kg body weight.

Most meat chickens are reared entirely indoors in environment-controlled sheds (Figure 7-5), featuring automated feeders and drinkers, artificial

Figure 7-5. Typical meat chicken housing.

lights and littered floors. They generally live in large, single-age flocks with high stocking densities, typically 33–39 kg/m^2 in EU countries. Some farms provide windows for natural light, now a requirement under the Red Tractor farm assurance scheme, which covers most commercial enterprises in the UK. However, natural light remains 'supplementary', and few sheds rely on it entirely.

Common welfare hazards

- Feeding regime.
- Genetic selection and management for high production.
- Stockperson to bird ratio.
- Stocking density.
- Group sizes (non-cage laying hens, meat chickens).
- Litter quality (non-cage laying hens, meat chickens).
- Environmental variety (caged laying hens).
- Degree of confinement and cage design (caged laying hens).
- Mutilations – beak trimming (laying hens).
- Staff competence and training.

Welfare hazards are the resources (inputs) that can lead to poor welfare outcomes. The likelihood of poor welfare outcomes and their severity (i.e. the risks) can vary. Severity encompasses both duration and intensity. Common welfare outcomes are identified below, structured by the Five Freedoms framework.

Chicken welfare is determined by how well chickens can adapt, both physically and mentally, to their environment (see p. 8). Relevant descriptions of 'free-living' chickens are provided throughout to highlight the adjustments they must make while in human care. Here, 'free-living' is defined as chickens living freely in an environment with the resources for which they have evolved and making their own choices. However, such comparisons provide only part of the welfare picture; welfare science (see Chapter 2) is essential for a proper understanding of the welfare impact of different husbandry systems.

Key determinants of welfare are the knowledge, skill and motivations of stockpersons, as well as whether they have sufficient resources (see pp. 24–25). Where these factors are optimal, welfare will be as good as it can be within a given husbandry system. However, there are often persistent welfare issues that are challenging to resolve without altering the husbandry system itself.

Common welfare outcomes

F1. Freedom from hunger and thirst

Chickens require a sufficient quantity and quality of feed to meet their nutritional needs and feel full after eating, as well as continuous access to clean water. These can be mostly provided in well-managed commercial farming systems. Generally, farmed chickens are provided with feed specifically formulated to support their high production. Feed is usually supplied in a pelleted or mash format and is offered *ad libitum*.

However, there are some challenges. Some welfare issues relate to management. For instance, for feed to be truly *ad libitum*, there must be sufficient trough space for all chickens. Ideally, space calculations will account for chickens having a strong drive to synchronise their behaviours. Studies of groups of laying hens indicate they are motivated to synchronise their main feeding bouts, which occur early in the morning and late in the afternoon. If quality, quantity or access are limited, subordinate chickens are likely to experience hunger due to competition from more dominant birds.

Meat chickens have been less studied than laying hens, but current evidence suggests some important behavioural differences. Excess competition in laying hens is more likely to lead to aggressive encounters. Meat chickens generally show little threat or aggression, even when they cluster around highly valued resources, such as feeders.

Behavioural differences between laying hens and meat chickens are unsurprising, as they are distinct breeds and strains of birds, and very different ages – meat chickens are juveniles, typically killed by 6–8 weeks of age, whereas laying hens are adult birds. (Note, however, that the parent birds of meat chickens, i.e. broiler breeders, can exhibit considerable aggression.)

There are also inherent welfare issues, i.e. even where farming systems are optimally managed, with stockmanship as good as it can possibly be, certain key welfare issues can remain. Again, key issues differ when comparing laying hens and meat chickens. Examples are as follows:

Osteoporosis in laying hens

In laying hens, genetic selection for high production can cause nutritional problems. Most farmed laying hens are under considerable metabolic stress, and the diet often fails to meet the high demand for calcium required by egg production, which can lead to osteoporosis and associated problems (see F3).

Reduced mobility in meat chickens

Meat chickens have been bred and managed to develop a large breast muscle mass, which significantly alters their morphology compared with 'free living' chickens. This can reduce their mobility later in the rearing period (see F5), and they may develop limb pathologies (see F3), both of which can impede birds' access to feed and water points.

Chronic hunger in broiler breeders

Note that whilst meat chickens are fed *ad libitum*, the parent birds (broiler breeders) commonly experience chronic hunger, as they are commonly offered only around one-third to one-half of their preferred food intake. This is because genetic selection for rapid growth has given broiler breeders an overly large appetite, to the point that if they are fed a standard diet to appetite, they will become overweight and experience a multitude of health problems, including skeletal and metabolic disorders, and increased mortality. However, alternative, more welfare-friendly strategies could be employed, such as providing a lower-calorie-density feed, which allows broiler breeders to feed to satiety with a much reduced risk of health problems.

F2. Freedom from discomfort

To be comfortable, chickens require sufficient quantity and quality of rest. They require comfortable resting surfaces, as well as sufficient space to exhibit basic comfort behaviours, such as stretching their legs and wing flapping. They must avoid being too hot or too cold. They benefit from engaging in basic body care, such as preening and dust bathing (see F5).

Providing space and resources for all the chickens in the group to be comfortable can be challenging in the main commercial farming systems. There are multiple interrelated factors to consider, and again, the main issues vary when comparing meat chickens with laying hens, as well as laying hens in cage and non-cage systems. Examples are as follows:

Thermal comfort

Thermal comfort is very important for both welfare and productivity in all chickens.

There is generally more scope to provide chickens with consistent thermal comfort when they are housed indoors, as they can be protected from inclement weather, particularly cold stress. Indeed, this is one of the reasons indoor farming systems were originally developed.

However, excellent management is needed for the benefits to be realised. Even minor deviations in parameters such as air temperature, ventilation and humidity can significantly alter the temperature the chickens experience. Heat stress can be a significant issue for laying hens and meat chickens housed indoors, particularly at high stocking densities, as they have limited behavioural means to lose body heat. Heat stress, can escalate, causing high morbidity and mortality (see F3).

Resting behaviour

All chickens require comfortable resting areas and sufficient uninterrupted resting time. There are some differences when comparing farmed laying hens in cage and non-cage systems, as well as when comparing laying hens with meat chickens.

Risks for laying hens in different systems

Laying hens in non-cage systems can express a much wider range of behaviours than those in cage systems. The extent to which they achieve comfort will depend on how well the system is managed. Non-caged birds usually have space to stretch their legs, wing flap and preen. In well-managed systems, they will often have access to dry friable floor bedding ('litter') that allows them to dustbathe. They are usually provided with perches, and there are other elevated areas where they may choose to rest. Resting and sleeping at a height (i.e. perching and 'roosting') are examples of highly motivated behaviours that evolved to protect against predation (see F4). Perching and roosting likely improve the general comfort of birds as well as their feelings of security. Factors such as perch height, diameter, shape and space per bird will all impact the degree of comfort achieved. Preferred perch parameters have been extensively studied in laying hens. In contrast, laying hens in conventional cages have all their behaviours severely constrained.[90] 'Enriched' cages provide laying hens with more space and furniture than conventional cages, but performing basic comfort behaviours can remain difficult.

Risks for meat chickens

Meat chickens have been selectively bred for a large breast muscle mass (see F3) and have a very different morphology from laying hens. This may partly explain why they spend longer periods of time resting by lying down on the floor, particularly later in their growing period. Their skin is in contact with the floor surface during these extended lying periods. Poor litter (bedding) quality is a common cause of discomfort, which can escalate to cause pain, injury and disease (see F3). To mitigate, multiple interrelated factors must be well-managed to maintain a dry, friable litter quality. Factors include ventilation, humidity, stocking density, feed type (as this affects the consistency of bird droppings) and drinker design.

The resting behaviour of meat chickens is often disturbed by other birds walking over them on their way to the feeders and drinkers, in contrast to laying hens, which tend to walk around each other. This likely explains why meat chickens prefer to rest close to the building walls and other vertical barriers, as they are less likely to be disturbed in these areas.

In addition, young broiler chickens may sometimes be exposed to extended light patterns for production reasons, which further disrupts any behavioural synchrony, resulting in greater disturbance of resting birds by active ones. Some countries limit uninterrupted darkness; for example, there are EU laws that mandate minimum periods.

Generally, meat chickens have room to stretch their limbs and flap their wings for much of the rearing period. There is usually no or very limited perch provision (see F5).

F3. Freedom from pain, injury and disease

Commercial strains of chickens with high production potential require careful management to thrive, whether laying hens or meat chickens. When conditions are optimal, performance can be very good, and the system's welfare potential can be met. However, even small deviations in parameters such as nutrition, temperature, ventilation and humidity can lead to a rapid deterioration in health, welfare and production. For example, heat stress (see F2) can easily escalate in farmed chickens. A vicious cycle can develop where the birds that are feeling too hot pant to lose heat, and the muscular effort required to pant generates further heat, escalating to heat exhaustion and death. Even relatively modest rises in ambient temperature pose risks if humidity levels are high.

Shortcomings in husbandry can predispose to non-infectious diseases and injury. These, in turn, can predispose to infectious diseases. Some very contagious, infectious diseases establish themselves even where birds have good prior fitness levels. Infectious diseases can spread more rapidly where large numbers of chickens are housed at high stocking densities. Mitigations are essential.

Good biosecurity measures can help prevent disease-causing organisms from entering the site via workers, visitors, shared equipment and small wild animals, such as rats and birds. Some diseases can be transmitted from parents to offspring. Therefore, an 'all-in-all-out', single-age production system is desirable, preventing older contaminated birds from spreading disease to young naïve birds arriving on site.

The common sources of pain, injury and disease differ when comparing laying hens in cage and non-cage systems, as well as when comparing laying hens and meat chickens. Laying hens in non-cage systems tend to be at a greater risk of suffering poor health and high mortality than their caged counterparts.[90] Feather pecking and smothering are also common in non-caged laying hens. The key welfare issue for meat chickens is that they can be affected by painful conditions that cause lameness.[90]

Particular issues in laying hens

Risks of infectious disease vary by system. Morbidity and mortality rates of laying hens are generally higher in non-cage systems, because many diseases spread via the faecal-oral route. Cage systems have better hygiene because hens are completely separated from their droppings. Additionally, maintaining biosecurity is more challenging when chickens have access to the outdoors.

However, there are welfare trade-offs, as multiple welfare benefits are afforded by the greater behavioural expression enabled by non-cage systems and access to a well-managed range (see F5).

Health risks can increase or decrease according to management levels. For instance, if an outdoor range is poorly managed, with insufficient aerial cover, birds will be afraid to venture far from the house. They gather near the 'popholes', creating muddy areas with considerable faecal contamination, and readily ingest pathogens, particularly endoparasites, when they forage here. Outdoor birds are more susceptible to diseases carried by wild bird populations, such as avian influenza. They are also more vulnerable to predation by wild animals, such as foxes.

Smothering in non-cage systems (laying hens)

Chickens are a prey species, evolved to remain vigilant and to synchronise their behaviours with other flock members, as this improves their survival chances. Their fear-driven escape behaviours aid survival in their natural environments but can be detrimental to welfare when performed synchronously by large groups of chickens in the confined spaces of farming systems. Smothering can lead to significant suffering and mortality in laying hens in non-cage systems, where birds that flee tightly aggregate and pile on top of each other. Something as simple as a person suddenly and unexpectedly entering their house or a sudden noise can cause mass alarm. To mitigate this, chickens can be habituated to different sights and sounds, particularly at a young age, and stockpersons can ensure they behave in a calm, quiet and predictable manner.

Generally, smothering is less of an issue in meat chickens as they tend to be less flighty than laying hens. However, smothering can occur on occasion. Factors other than flightiness can contribute, such as birds tightly aggregating when they are simultaneously attracted to a preferred resource, such as a particular area of lighting.

Bone fractures (laying hens)

Egg laying is primarily driven by the birds' natural photoperiod, and places all chickens under considerable metabolic stress, whether they are farmed or 'free-living' chickens. In all cases, their welfare will be impacted by the extent to which the quality of their diet meets their needs.

Factors that can help mitigate metabolic issues in 'free-living' chickens include their seasonal laying pattern, characterised by a combination of clutch size and 'broody' behaviour, which intermittently pauses egg laying during this period. In contrast, most commercial strains of laying hens lay daily for much of the year with little pause and have lost much of their broody behaviour. Even in well-managed farming systems, the diet can fail to meet the high demand for calcium for eggshell production (see F1). Laying hens then deplete their skeleton to meet this demand, leading to osteoporosis and associated problems, such as fractures.

Environmental conditions can exacerbate the injuries and pain caused by osteoporosis. Although hens housed in cages tend to have a greater degree of osteoporosis through disuse of body parts, injuries/fractures tend to be worse (more prevalent) in hens in non-cage systems. Non-caged laying hens remain active and sustain stronger bones than their caged counterparts, but they are more exposed to trauma. They commonly suffer fractures, even with only limited trauma. Over half of the birds in the flock can be expected to suffer keel bone fractures due to colliding with perches. Attention to housing design and furniture placement can help to mitigate this. When handled to empty the housing at the end of lay, laying hens from both cage and non-cage systems are at high risk of fractures, especially limb fractures.

Feather pecking (laying hens)

Feather pecking is an 'abnormal' behaviour displayed predominantly by laying hens in non-cage systems (see F5). It affects the majority of non-cage flocks by 40 weeks into the laying period. It can also occur in cage systems, but predominantly affects non-cage systems because the behaviour spreads much more readily where birds are kept in large groups. Pecked birds can suffer substantial plumage loss and skin damage, which increases their risk of infection and mortality, as well as causing pain, fear and distress. Outbreaks can be difficult to predict and hard to control, despite extensive research. Research indicates that feather pecking is multifactorial, with risk factors including high stocking densities, fearful birds, dietary alterations, feed format and genetic selection for high production. However, a key risk factor is the lack of suitable litter or other materials for pecking and scratching. Hens are thought to then redirect their foraging behaviour at other hens (see F5).

Beak trimming (laying hens)

Beak trimming is a mutilation often performed routinely on day-old chicks destined for laying systems. It is done to try to reduce the severity of feather pecking in laying hens. The sensitive tip of the chick's beak is trimmed using either a hot blade or an infrared beam. Trimmed beaks remain more sensitive and less precise into adulthood, making it harder for the hens to peck. The birds experience acute pain at the time of the procedure and possibly longer-term pain due to neuroma formation at the trim site. While the procedure reduces the severity of injury to pecked birds, it does not address the underlying reasons why the birds peck others (see F5).

Laying hens can also be injured by having their vents or preen glands pecked, independently of any feather pecking.

Particular issues in meat chickens

Meat chickens have been selected for rapid growth and disproportionately large breast muscle, placing increased stress on their circulatory system and immature skeletons. This can cause multiple health problems, not experienced by laying hens.

Lameness is common in fast-growing strains of broilers from 3–4 weeks of age and generally worsens towards the end of the rearing period. Underlying limb pathologies include angular deformities, skeletal calcification issues such as tibial dyschondroplasia and bacterial infections of joints. Ulcerative skin conditions, predominantly footpad dermatitis (pododermatitis) and hock burn, are common. Inactivity exacerbates limb and skin disorders.

Poor quality litter further irritates the skin of birds. Multiple factors reduce litter quality, including high stocking rates, poor control of temperature, humidity, and feed types that cause greasy faeces. Studies demonstrate that many limb and foot disorders are painful; for example, gait improves when birds are administered analgesics.

F4. Freedom from fear and distress

Unpleasant feelings such as fear and anxiety often accompany pain, injury or disease. They can also occur independently, triggered by multiple factors.

Chickens are a flighty prey species that evolved to remain vigilant. Momentary fear and alarm are a normal part of everyday life and do not necessarily constitute welfare issues, especially if animals have the choice to take action to alleviate fear, such as escaping or hiding. However, excessive or prolonged fear and/or anxiety lowers their welfare.

Predation (real or perceived)

All farmed chickens, whether laying hens or meat chickens, can experience fear of predation. Chickens with outdoor access are more vulnerable to real predation by wild animals such as foxes. Chickens in any system may perceive themselves at risk of predation.

Chickens can perceive their human carers as potential predators. Calm, quiet, familiar and predictable movements by stockpersons are essential to mitigate. Opposite actions are likely to cause fear. Sudden, unfamiliar noises or sights can cause alarm. The fear-driven escape behaviours of large groups of chickens in a confined space can lead to 'smothering' (see F3).

Chickens are innately fearful of wide-open spaces. Therefore, the features of an outdoor range determine whether the chickens use it and how much it benefits their welfare. Overhead cover provided by trees, shrubs and artificial shelters encourages birds to range widely, as they feel safer.

To mitigate fear and anxiety, it is important that all chickens in a farming system can perform highly motivated behaviours that make them feel secure. A key behaviour in laying hens is roosting (perching at a height to rest and sleep) during the night. Farmed laying hens retain the same strong motivation as 'free-living' chickens to roost when the light fades in the evening, as in these locations, they are better protected from ground predators. If they ignored these cues in the natural environment, they would jeopardise their survival. They do not know that the farming system has provided them with manmade protections from real predators. If a lack of perch space prevents subordinate hens from perching, they are likely to be fearful as a result.

Social stress

Chickens have evolved to live in small, stable groups and synchronise their behaviours with those of other group members, strategies thought to reduce the risk of predation and improve their survival chances. Therefore, it is beneficial for the welfare of farmed chickens that they live in groups and experience social isolation only rarely. However, group dynamics are important and can be altered, for better or worse, by factors such as group size, space and the availability of resources.

The very large group sizes of non-cage systems can cause considerable social stress in laying hens. This is because chickens have evolved to live in groups of up to 30 individuals, forming a social hierarchy or 'pecking order', where they remember their relationship with every other bird. This social hierarchy supports social harmony and less aggression when accessing shared resources. Studies show they can form such hierarchies in group sizes of up to 100 individuals. However, they struggle to do this in large group sizes of hundreds, sometimes thousands, of birds and are likely to experience unpleasant feelings such as anxiety.

Laying hens in all systems need sufficient space and resources to prevent excess competition and the associated aggression, and to escape and hide when feeling threatened by other birds. Otherwise, they can become chronically fearful. Stable social hierarchies can form in the small group sizes of cage systems, reducing aggressive encounters. However, on the occasions when there is aggression, the chickens have nowhere to escape.

The behaviours and motivations of meat chickens have been far less studied than those of laying hens, and further research is needed to identify how they compare with laying hens.

F5. Freedom to express normal behaviour

Domestic chickens have retained many behaviours of their wild ancestor, the Red Jungle Fowl. It is simply the intensity with which such behaviours are performed that has altered.

The welfare importance of many laying hens' behaviours has been extensively studied. The limited studies of meat chicken behaviours suggest they differ from laying hens in certain ways.

Particular issues in laying hens

Findings show that the behaviours that laying hens are highly motivated to perform much of the time include foraging, exploring and social interactions with other chickens. A key, highly motivated behaviour performed at a specific time is nest building, which occurs prior to egg laying. Dustbathing and preening are examples of behaviours that they are motivated to perform for short time periods only, but are important to them nonetheless. Allowing for these behaviours can support good welfare. Thwarting them can lead to unpleasant emotions such as frustration and boredom.

The scope for normal behavioural expression in laying hens varies according to whether they are kept in cages or non-cage systems. In cage systems, their behaviours are severely constrained.[90] In conventional cages, even basic movements are difficult, and the environment is barren and unstimulating. Enriched cages are better than conventional cages, providing more space and furniture to allow for some nesting, perching and scratching; however, such behaviours are still very limited. Well-managed non-cage systems allow for much fuller behavioural expression. A well-managed range with sufficient cover (see F4) offers further choice and interest. However, there are still deficiencies in the birds' environment, as evidenced by the high prevalence of 'feather pecking' in non-cage systems.

Understanding the importance of these behaviours to laying hen welfare involves understanding the importance and function of these behaviours in 'free-living' chickens:

Foraging and exploring

Hunger and thirst can be largely avoided in commercial farming systems (see F1), but the behavioural aspect of eating may be thwarted, causing unpleasant feelings such as frustration. This is primarily an issue for caged laying hens.

'Free-living' chickens are omnivores that consume a varied diet and have a keen ability to detect small, moving, invertebrate prey items. They are highly motivated to forage and explore for much of the day, largely by pecking and scratching. Studies show they will choose to 'work' to find food, even when food is freely available. When small flocks of domestic chickens are kept in semi-natural environments but fed on commercial feed rations, they can meet their nutritional needs by eating for only 1–2 hours daily, yet still choose to explore their surroundings by pecking and scratching for the remaining daylight hours.

Feather pecking (laying hens)

Feather pecking is an 'abnormal' behaviour, predominantly of caged laying hens (see F3). Pecked birds suffer injury, pain, fear and distress. Abnormal behaviours generally indicate that the animals' environment lacks something they need i.e. the birds that peck also have suboptimal welfare.

Feather pecking is multifactorial. However, it is thought to be partly a frustrated, redirected foraging behaviour where good-quality litter or other suitable materials for pecking and scratching are lacking. Farmed birds quickly consume their commercial feed rations and are left with a lot of free time and little to do. Here, some birds begin to peck at the feathers of other birds, even pulling out and eating them. Feather pecking remains a challenging behaviour to control, despite extensive research. Management methods, such as beak trimming (see F3) and reducing light intensity, cause their own welfare problems.

Social interactions

Maintaining harmonious social interactions and low social stress in farming systems can be challenging due to the altered group sizes and environment compared with those of 'free-living' chickens (see F4).

Nest building (laying hens)

Farmed laying hens retain a strong, hormonally driven need to build a nest at the time of egg laying. Studies show this elaborate behaviour will occur even if a pre-made nest is available. Around one hour before laying, hens become restless and seek solitude and a nest site. Nesting typically begins around half an hour before laying, although it can take significantly longer in some birds. Enclosed sites with manipulable materials are preferred. The hen alternates between sitting and building activities, including scratching the floor and collecting nearby pieces of straw or materials that she places around her body and tosses over her back. The ability to mould a nest using body and foot movements seems to be particularly important.

In conventional cages, none of this is possible. In enriched cages, nest areas provide a darkened and secluded laying area, preferable to a barren cage but lacking space and materials. In contrast, non-cage systems provide the space, nest boxes, friable floor materials and a greater choice of areas. While vastly better than cage systems, the nesting behaviour is still limited compared to that of 'free-living' chickens.

The importance of roosting and perching behaviours has already been highlighted under F4.

Dustbathing and preening

'Free-living' chickens perform comfort and body care behaviours such as preening and dustbathing. These two behaviours are performed for short periods of time but remain important for the bird – studies show that if laying hens are denied these behaviours, there will often be a 'rebound effect' with excessive performance when the behaviours are finally allowed.

These behaviours serve different purposes. For example, when birds preen, they clean and align their feathers with the beak, whilst distributing oil from the uropygial gland. Dustbathing removes lipids and ectoparasites from feathers and is performed for a few minutes daily. Here, the chickens rake loose substrate on the floor with their beaks before lying down, turning slightly to one side, raising their feathers, shaking their wings, rubbing their heads, and scratching with their legs. They then stand up and vigorously shake their body to remove the loose substrate. Sand substrate is preferred over straw. However, these behaviours are severely limited in caged birds. There is much greater expression in non-cage systems.

Despite the extensive scientific study of laying hen welfare, there is still much to learn about their communication, cognition, learning and problem-solving abilities, including how to apply these findings practically. One area less understood is the welfare impact of single-age groups. Farmed chickens are reared and kept in single-age flocks for biosecurity and production reasons (see F3), which is very different to the multigenerational flock of 'free-living' chickens, where chickens are exposed to the adult environment from hatching and are likely to learn much from their mothers and other birds in the flock, following them closely and imitating their actions. For example, when farmed laying hens are placed in layer accommodation, they will not automatically want to use the nest boxes provided. They develop their nest site preferences before sexual maturity and are unlikely to change them, even when later provided with new locations with facilities that should be preferable. With creative thinking and attention to detail, mitigations can be implemented that may improve both bird welfare and production. Examples include introducing birds to nest boxes at 14–16 weeks of age before they start laying. This can also improve production by reducing the number of floor-laid eggs, which are more challenging to collect and are more likely to break and become dirty.

Particular issues in meat chickens

The limited studies of meat chicken behaviours show there are significant differences when compared with laying hens. This is unsurprising, given that broilers are juvenile birds, killed at around six weeks of age, and also belong to very different breeds, having been selectively bred for large breast muscle mass. The extent to which these differences in behaviour are due to differing motivations and/or due to altered physical ability to express certain behaviours needs to be established.

It is beneficial for the welfare of meat chickens that they are kept in groups and can interact with one another. In well-managed systems, they have access to good-quality litter for scratching, foraging and dust bathing. Initial observations suggest that there is very little overt competition, threats or aggression between meat chickens, even when they cluster around highly valued resources, such as dry litter, feeders and vertical barriers, such as walls.

However, their environment lacks complexity compared with that of noncaged laying hens, and they have few other choices for behavioural expression that could enhance their welfare. Having said that, meat chickens need resources tailored specifically for them, rather than extrapolated directly from laying hens. For instance, their physique appears to alter their activity compared with 'free-living' types and strains of chickens with their large breast muscle affecting their gait, centre of gravity and posture. Most meat chickens can move relatively freely during the first three weeks of life, but their activity then dramatically decreases, even in birds with minimal leg pathology. Those experiencing painful limb conditions (see F3) will have their activities further limited.

Studies show that meat chickens are motivated to perch, which is unsurprising – a 'free-living' chicken would start perching at around two weeks of age, developing these skills by following the mother to progressively higher roosts for a longer duration until settling into nighttime roosting at around six weeks old. However, perches for meat chickens should be located much closer to floor level. Meat chickens can make effective use of straw bales for many different purposes – perching, pecking, foraging, support when resting and vertical barriers that limit disturbance by other chickens when resting.

Some farms now provide these as enrichment. For instance, the Red Tractor farm assurance scheme, which covers most UK standard commercial production, requires at least one straw bale and one pecking object per 1000 birds, and at least two metres of perch length per 1000 birds. Further research could determine the enrichment levels needed for all chickens to have adequate access and address logistical challenges.

Ultimately, the root cause of many welfare issues in meat chickens needs to be addressed by transitioning to slower-growing breeds with a more balanced conformation.

Welfare assessment protocols

AssureWel[23] and Welfare Quality®[8] have developed protocols for laying hens and meat chickens that comprise mainly animal-based measures (see Chapter 3).

Advocacy examples

- The Laying Hen Welfare Forum[91] is a UK national initiative that brings together stakeholders from industry, farming, animal welfare and the government to improve laying hen welfare. For example, their FeatherWell project provides excellent resources on injurious feather pecking.
- The Better Chicken Commitment[92] is an international initiative that sets welfare standards for meat chickens, which food companies can choose to adopt voluntarily, demonstrating their commitment to animal welfare and reducing associated business risks. An example is the use of specific slow-growing breeds. To date, over 200 leading food companies, including several large multinational corporations, have embraced these standards.
- Legal minimum standards for chickens in EU countries are strongly informed by and intermittently amended according to scientific opinions of the EFSA Panel on Animal Health and Welfare. As a result, conventional cages for laying hens have been illegal in EU countries since 2012, and there are legal minimum standards for resources, such as space allowance, nesting and perching facilities, in other husbandry systems.[93] EU laws for meat chickens have introduced a requirement for the number of birds 'dead on arrival' at slaughterhouses, as well as certain post-mortem findings at slaughter, to be monitored. Findings are used to provide evidence of welfare on farm, and for actions to be taken accordingly, which is something of a novel, progressive approach in law.[94]
- Potential future advocacy could involve the parent flocks and their particular challenges. For example, consumers are often less aware of the degree of food restriction and the levels of aggression amongst broiler breeders.[90]

Cattle 8

Abstract
The core welfare concepts discussed in Part 1 are applied to cattle. Common welfare hazards are listed. Common welfare outcomes are identified and structured according to the Five Freedoms framework. Practical welfare assessment protocols are referenced. Examples are provided of how veterinary professionals have advocated for cattle or could be more involved.

Introduction

Cattle (*Bos taurus*) have been domesticated for thousands of years. Over that period, a synergistic bond has developed between humans and cattle, which, at its best, benefits both: cattle provide meat, milk, skins, traction and financial security, and, in return, they receive shelter, food and protection from predators. Today, cattle are farmed on every continent in a wide range of environments, from the icy regions of Canada and Mongolia to the arid plains of sub-Saharan Africa.

Cattle farming systems are equally varied. Life in some communities continues to revolve around cattle to this day, for example, the Maasai peoples of Tanzania and Kenya. These are often nomadic, rural groups who manage their herds in traditional ways that have changed little over hundreds of years, though problems of land tenure may alter access to grazing areas. At the other extreme are the urbanised societies, where the proportion of the population engaged in cattle farming has declined dramatically over recent decades. At the same time, per capita consumption of beef and dairy products has increased. This has necessitated the development of new farming systems that enable very large numbers of cattle to be managed by small numbers of people. Genetic selection, better nutrition and improved understanding of disease control have enormously increased production per head, with dairy cows now producing in excess of 13,000 litres of milk per lactation and 60 litres per day, and beef animals growing at up to 2 kg bodyweight daily. There can be tension between high production driven by economic constraints and livestock welfare.

DOI: 10.1079/9781786391582.0008

This chapter illustrates core welfare concepts with examples from beef and dairy cattle farming predominantly in European countries.

Husbandry

Cattle farming in European countries can be broadly categorised into dairy and beef farming. Husbandry practices vary significantly between and within sectors, influenced by factors such as geography, climate, feed availability and market demands. Systems range from extensive 'low input, low output' to intensive 'high input, high output'. Some features commonly seen are summarised below.

Dairy farming

The primary role of dairy cows is to produce milk. They typically give birth to one calf each year and then lactate for around 10 months. They have a non-lactating (dry) period of around 2 months, before giving birth again, having been mated or artificially inseminated at 3 months post-partum. Gestation is 9 months.

Dairy cows are generally milked mechanically in a milking parlour 2–3 times daily (Figure 8-1), though the use of robotic milking systems, which allow cows to enter a unit and be milked automatically without human intervention whenever they choose, is increasing.

Housing and access to the outdoors vary. Generally, cows are loose-housed in groups, though some small farms in some countries still use 'tie-stalls'. Some

Figure 8-1. Dairy cows being milked in a herringbone parlour.

farms permanently house the animals. Others provide access to outdoor areas (loafing areas or pasture). Pasture access is often seasonal.

Housing designs vary but can be broadly categorised into cubicle ('free-stall') housing and open-bedded yards. Cubicles are individual lying areas within the housing that define resting areas for individual cows (Figure 8-2) and were developed to ease the management of larger herds, for example, providing cleaner resting areas and facilitating manure collection. Each cubicle has a raised concrete base and partitions, including a neck rail, front rail, and brisket board, to guide the cow into the correct lying position. Floors may be covered with synthetic mattresses or bedded with materials such as straw, sawdust or sand. There are passageways between rows of cubicles and a feed alley, providing space for cows to move, loaf and feed. Excreta are deposited on passageway floors, which are cleaned using various systems such as manual or automatic scrapers.

Straw-bedded, open yards may be used alongside cubicle housing or may be the sole housing. Pens are generally deep littered with fresh straw added in layers daily. Often, dry or recently calved cows are housed in open-bedded yards while lactating cows are housed in cubicles. Loafing areas are additional spaces, usually located adjacent to the housing, that offer opportunities for exercise and access to fresh air and natural daylight.

Access to pasture for grazing varies with management and season. Many farms combine pasture-based production during the summer with indoor housing in the winter, though some herds are housed year-round.

Dairy calves are typically separated from their mothers at or shortly after birth and artificially reared, being fed milk or milk replacer until they are about

Figure 8-2. Dairy cow resting in a cubicle.

Figure 8-3. Artificial rearing of calves from the dairy system.

6–9 weeks old (Figure 8-3), which is much earlier than natural weaning. After this, calves are reared on solid foods, such as concentrates and forages, either on their birth farm or in specialised calf-rearing units. Some female dairy calves remain within the dairy system, being reared to replace cows that have reached the end of their productive life. Other dairy calves (male and female) enter beef production systems.

Beef farming

Husbandry systems for beef cattle can be broadly categorised as indoor or grass-based. Many beef cattle are reared entirely indoors in intensive fattening units. Typically, they are housed in groups in straw-bedded barns that are fully or partially closed. They are typically fed significant cereal-based (concentrate) feed in addition to forage to increase their rate of growth (Figure 8-4), and are slaughtered at 12–16 months old.

Fattening cattle originate either from the dairy system (see above) or from beef suckler herds. In beef suckler herds, the primary role of adult cows is to produce calves that will be reared for meat. Breeding herds are usually kept outdoors at pasture, where their diet is predominantly forage from grazing (Figure 8-5). At times, they are offered supplementary forage as conserved grass. Concentrate feeding is minimal. Breeding is typically achieved through natural service using a bull. Beef suckler calves remain in the herd alongside their mothers, suckling them until they are separated at around 5–9 months old, after which they are reared separately on various diets. Many go to intensive fattening systems. Others remain in extensive outdoor production

Figure 8-4. Beef cattle being fattened in indoor housing.

Figure 8-5. Beef suckler cattle at pasture.

systems, with pasture access available during summer or year-round, and grow at a slower rate than in intensive systems. Those extensively reared are typically slaughtered between 16 months and 2.5 years, depending on the feeding system and breed.

Common welfare hazards

- Feeding regime.
- Design and maintenance of housing.
- Pasture management, shelter and shade.
- Exposure to infectious disease.
- Genetic selection for production characteristics.
- Quality of stockmanship.

Welfare hazards are the resources (inputs) that can lead to poor welfare outcomes. The likelihood of poor welfare outcomes and their severity (i.e. the risks) can vary. Severity encompasses both duration and intensity. Common welfare outcomes are identified below, structured by the Five Freedoms framework.

Cattle welfare is determined by how well cattle can adapt, both physically and mentally, to their environment (see p. 8). Relevant descriptions of 'free-living' cattle are provided throughout to highlight the adjustments they must make while in human care. Here, 'free-living' is defined as cattle living freely in an environment with the resources for which they have evolved and making their own choices. However, such comparisons provide only part of the welfare picture; welfare science (see Chapter 2) is essential for a proper understanding of the welfare impact of different husbandry systems.

Key determinants of welfare are the knowledge, skill and motivations of stockpersons, as well as whether they have sufficient resources (see pp. 24–25). Where these factors are optimal, welfare will be as good as it can be within a given husbandry system. However, there are often persistent welfare issues that are challenging to resolve without altering the husbandry system itself.

Quality of stockmanship

The actions of stockpersons directly impact all aspects of animal welfare, and specific examples are provided under each 'Freedom'. However, addressing failures of individual stockpersons is challenging without also addressing the important root causes.

Poor stockmanship is often caused by failures at the management level. Management may recruit unskilled staff, fail to train them adequately and/or fail to provide and support the implementation of health and welfare plans, including Standard Operating Procedures (SOPs) for husbandry tasks. Poor communication exacerbates problems. While poor empathy with animals may be an innate characteristic of some stockpersons, it is also influenced by the example set by more senior and experienced workers.

Even in very competent stockpersons, a lack of attention to detail and failures of observation can result from an excessive workload, with workers being too busy, having responsibility for too many animals, or working excessively long hours. This problem often escalates as poor hygiene, improper feeding and other factors raise disease levels, further increasing the workload; one sick cow can require the care and resources of several healthy ones.

All aspects of farming can be adversely affected by poverty or lack of financial investment. Prices paid for cattle produce have a major influence on profit potential and the level of investment farmers can afford. If prices paid to farmers for cattle produce are driven down by supermarkets and consumers who demand cheap food, animal welfare is likely to suffer. This can also become an escalating problem: a lack of investment leads to poor nutrition, dilapidated housing and a failure to implement disease prevention measures, such as vaccination and the use of anthelmintics. This results in increased incidence of disease and/or reduced productivity. In turn, lower profits result in less money available for investment in cattle farming.

Common welfare outcomes

F1. Freedom from hunger and thirst

Cattle are ruminants, evolved to feed on forage, which they obtain by grazing for much of the day. 'Free-living' cattle usually have two main grazing periods, one in the early morning and another in the late evening or around dusk, though they may graze more at night when the weather is hot. They are social animals that live in stable groups with an established social hierarchy that minimises confrontations between animals. They do not require much sleep, but they need sufficient undisturbed rest time (around 8 hours) to ruminate. Mothers suckle their calves until the milk supply declines as the next pregnancy advances into the last trimester, and weaning is a gradual process.

Imbalanced nutrition is a main risk for farmed cattle. They have higher metabolic and nutritional demands than their free-living counterparts, as they are used to produce meat and milk and generally have a larger body size.

In extensive farming, herds rely mainly on outdoor grazing year-round, and this most closely resembles the feeding of 'free-living' cattle. However, they often require supplementary feeding at key times, for example, to compensate for seasonal decline in grass growth. Natural events, such as severe weather, and management practices, like overgrazing, can reduce the quality and quantity of grazing. Even good-quality grazing may not adequately meet the nutritional demands of high production. In extensive systems, supplementary feed is often conserved grasses, i.e. straw, hay and silage, which cattle are well adapted to consume and digest. However, inadequate feed provision can cause hunger. A common cause is where farmers struggle to meet the financial costs of feed, both for outwintered cattle and those housed over winter. Subordinate animals tend to be disproportionately affected by competition within the herd.

Forage alone is insufficient to meet the nutritional requirements of many highly productive cattle, and they often require supplementation with cereal-based feeds to avoid energy shortfalls. For instance, the nutritional requirements of high-producing dairy cattle have been compared with those of a professional athlete or top-class racehorse. However, dairy cattle are managed in large groups, making it challenging to meet their individual needs. Metabolic diseases commonly result from such high production demands. Water requirements for cattle are also high, particularly for lactating dairy cows. A dairy cow producing 40 litres of milk per day requires around 100 litres of water per day.

Every aspect of nutrition must be correct, from identifying nutritional requirements and ration formulation through to delivering the correct ration, monitoring cattle performance and intervening if problems arise. All this requires technical knowledge, excellent stockmanship and adequate financial investment. For example, feed rations can be correctly formulated for high-producing dairy cows, despite the difficulties in meeting nutritional demands; however, it remains the responsibility of the stockperson to ensure that cattle receive that ration fresh every day and that plenty of water is available at all times. Failure of the stockperson to recognise the early signs of nutrition-related problems, such as weight loss, poor appetite and diarrhoea, can lead to their escalation and disease.

F2. Freedom from discomfort

In a natural environment, cattle express a range of behaviours important for their comfort. They select natural sources of shelter and shade to protect themselves from adverse weather and insects that irritate them. They seek ground that is clean, dry and deformable on which to lie and rest for extended periods while ruminating. They have room to stretch and adopt preferred postures. They engage in body care behaviours, such as grooming themselves, grooming others, and rubbing their bodies against trees and other suitable objects, to maintain coat condition and alleviate itch.

The comfort of farmed cattle depends on the type, quantity and quality of resources provided. Cattle are susceptible to heat stress and may also suffer cold stress, particularly young calves. For cattle grazing at pasture, the risks vary with weather conditions and the availability of shelter and shade, whether natural or man-made. Important animal factors include body size, coat thickness, fat coverage, and the level of acclimatisation. It is important to choose breeds/types that are well-suited to the available environment. Careful management is needed to limit poaching, which creates wet, muddy areas that are uncomfortable to lie or stand on.

Cattle are often housed during winter to protect them from weather extremes, facilitate management and prevent poaching of pasture. However, heat or cold stress may still occur if housing is inadequate — for example, if it is poorly ventilated, overstocked, draughty or insufficiently insulated and damp. Under such conditions, air quality can deteriorate, contributing to further discomfort. Normal movements are often limited by high stocking densities and a lack of space. This restriction makes it challenging for them to lie down in their preferred positions, get up after resting, and engage in other comfort behaviours, particularly when floors are hard or slippery. Behavioural synchrony, which cattle are highly motivated to achieve, can be challenging under such conditions. One common challenge in managing their environment is ensuring that straw bedding stays clean and dry.

Poor cubicle design and limited access are common causes of discomfort in dairy cattle. Housing that was effective in the past may no longer be suitable. For example, herd sizes may increase, but the available buildings remain the same, leading to overstocking. Selective breeding has generally increased cow size, and many cubicles are now too small to accommodate them adequately. Cattle that must stand excessively due to short cubicles are likely to be uncomfortable, as are those with poorly designed cubicle fittings and rough or slippery floors. Conversely, providing suitably thick, clean bedding or mattresses can significantly improve comfort.

Poor handling also causes discomfort. Examples are hurrying cattle along stony tracks or across concrete, and poor milking technique in the parlour, such as rough handling of teats and over-milking.

F3. Freedom from pain, injury and disease

Injury and disease are commonly associated with pain and other unpleasant feelings, such as fear and anxiety.

Reducing pain, injury and disease often improves production and farm profitability, making it financially sensible to do so. For instance, financial losses occur if disease causes deaths or reduces commercial outputs from the herd, such as growth, milk production and/or fertility. Therefore, disease control and prevention are important at the farm, national and international levels, even where there is little consideration for cattle welfare. There have been notable successes: national and international disease control strategies have eradicated certain infectious diseases in cattle, such as brucellosis in the UK and rinderpest globally.

However, despite extensive scientific research and the dedicated efforts of farmers and veterinary professionals, disease remains prevalent in farmed cattle. Notable examples include mastitis in dairy cows, lameness, metabolic disorders, respiratory illnesses and calf diarrhoea – all well documented in leading clinical texts. These conditions persist largely because many cattle diseases are multifactorial, with risk factors deeply embedded in modern farming systems. Elements such as large herd sizes, high stocking densities, suboptimal housing design and the physiological strain of high-yield production contribute significantly. While increased productivity is often pursued as a marker of progress, it does not always align with improved animal welfare. Lameness results from a combination of non-infectious and infectious factors – walking on concrete surfaces, inappropriate housing, large group sizes, poor hygiene and large body size. Genetic selection that focuses on production characteristics can significantly contribute to pain, injury and disease. Selecting dairy cattle for milk production has increased their susceptibility to lameness due to the additional metabolic stress and the excessive strain on the feet and legs caused by increased body weight. In beef cattle, genetic selection for extra muscle mass ('hypermuscling') predisposes them to calving difficulties and dystocia due to the mismatch between calf size and pelvic conformation of the dam. In some breeds, the extra muscling is so extreme that elective caesarean section surgery is routinely required.

Many injuries in farmed cattle are sporadic, but these, too, may be linked to husbandry factors, such as poor housing conditions. For example, slippery floor surfaces often cause animals to fall.

Mutilations, defined as procedures that interfere with sensitive tissues or bone structures performed for purposes other than medical treatment (see p. 20), cause pain and injury. Common mutilations in farmed cattle are disbudding, dehorning and castration.

Most breeds of cattle are naturally horned. Horns are commonly removed to allow for higher stocking rates and prevent injuries to other animals and humans. Disbudding involves removing the horn buds, which are located in the skin above the skull until the calf is around 2 months of age. Dehorning involves the surgical removal of fully grown horns in older animals after they have attached to the skull. The horns are supplied with nerves and blood vessels and remain sensitive throughout the animal's life. Castration involves the physical removal or inactivation of the testes for various reasons, such as easier management by reducing aggressive and sexual behaviours, and improving meat quality.

There are various methods for disbudding, dehorning and castration, and their welfare impact has been extensively researched. There are some differences when comparing techniques. However, they all cause distress and pain during restraint, procedure performance and the post-operative healing period. On farms, factors such as practicality, national laws and farm assurance scheme requirements determine the age at which procedures are carried out and the methods used. To protect welfare, effective anaesthesia and analgesia should be used during the procedure. Ideally, it will also cover the post-operative healing period, though this remains particularly challenging to deliver in practice. Certainly, welfare can be improved by basic post-operative nursing care, such as ensuring animals have comfortable, clean, dry beds.

Some breeds (mainly those used for beef) are naturally polled, and breeding practices can aim to produce more genetically polled types, removing the need for disbudding and dehorning. However, this is not straightforward, as genetic changes linked to polling can negatively impact production traits, particularly in dairy cattle.

Examples of how stockmanship failures lead to pain, injury and disease include:

- Failure to identify early signs of disease, when treatment is most likely to be effective and before infection has spread to others in the group.
- Inadequate implementation of disease control and prevention measures.
- Inadequate or incorrect treatment strategies so that recovery is delayed.

- Failure to provide care and nursing for sick animals that would minimise the welfare effects of disease.
- Poor hygiene, nutrition and milking routine, increasing the risk of disease.
- Impatient or unskilful interventions causing calving injuries.
- Failure of observation so that dystocia is not recognised early.
- Lack of attention to detail so that hazards (broken fencing, torn corrugated iron, damaged handling facilities) are left unrepaired and cause injuries.
- Impatient handling, causing cattle to panic and causing injuries.
- Failure to use adequate local anaesthetics and analgesia when painful procedures are performed.

F4. Freedom from fear and distress

Unpleasant feelings, such as fear and distress, are often associated with pain, injury and disease. They can also occur independently. Occasional fear and distress are an inevitable part of life, but they significantly impact welfare if they occur frequently or are intense.

Cattle are a prey species that generally evolved to be flighty in response to threats, real or perceived (see Chapter 17). These can come from humans or other animals.

Human–animal interactions

Unskilful handling and aversive stockperson behaviour, such as hitting, kicking and shouting, cause considerable fear and distress. Conversely, cattle can learn to relax in the presence of people, provided their experiences are positive, handling is calm and gentle, and they are socialised from a young age. This ability to learn (see pp. 20-24) is particularly evident in dairy cattle. In their daily routine, they become so familiar with people that their flight zone (see pp. 274-275) almost disappears in familiar, non-threatening situations. For example, they allow handlers to attach clusters to their udders when milked. The sound of the milking parlour, which differs from the sounds in their natural habitat, becomes familiar and a cue for milk let-down. They habituate to other unchanging features of their environment and will readily walk over contrasting surfaces, such as drain covers, which might initially cause alarm in a new setting.

Farmed cattle thrive on familiar routines, so changes, such as those encountered during transport or slaughter, can lead to significant fear and distress (see Chapter 17).

Interactions with other animals

'Free-living' cattle live in stable groups with a well-established hierarchy. Direct confrontations are uncommon. However, husbandry practices that disrupt these social dynamics and lead to more unfriendly interactions can induce fear and general stress, even in the absence of direct physical conflict. Situations that create excessive competition for vital resources, such as inadequate trough space that makes it difficult for cattle to access feed, can create considerable stress. Mixing unfamiliar animals disrupts social hierarchies, and aggressive encounters can occur as new hierarchies are established, creating considerable stress. For instance, young beef cattle are often mixed at weaning to form homogeneous groups based on specific characteristics. Careful management is needed to prevent escalation to fights and injuries. Cattle in very large herds can struggle to form a social hierarchy, as it becomes difficult for them to recognise one another when the group exceeds 100 individuals. This can lead to more unfriendly interactions and general increased stress levels within the herd.

In a natural environment, cattle are seldom or never alone. Social isolation causes intense fear. Mitigations include minimising the time and degree of separation of animals from their group mates when they are handled or placed in hospital pens, ensuring they can at least see, hear and smell other animals.

In 'free-living' cattle, the mother suckles her calf until the milk supply declines as the next pregnancy advances into the last trimester. Removing calves from their mothers shortly after birth – either immediately or within a few days – is a common practice in dairy farming. This separation can cause the mother acute distress that may last for several days, as they tend to form strong emotional bonds with, and have a strong hormonal drive to care for, their offspring and exhibit maternal behaviours. In contrast, beef suckler cows have the opportunity to develop strong bonds with their calves, which contribute to the well-being of both, with offspring weaned at a more natural pace.

F5. Freedom to express normal behaviour

Cattle retain and are highly motivated to express many of the behaviours of their wild ancestors. It is simply the threshold at which they occur and their intensity which have changed. Allowing for normal behavioural expression can support good welfare. Thwarting highly motivated behaviours can lead to unpleasant feelings such as boredom and frustration.

Some highly motivated normal behaviours have already been identified under the other Freedoms – grazing, ruminating, group living, behavioural synchrony, comfort behaviours and maternal behaviours. Additional important normal behaviours include friendly social interactions with other cattle, such as mutual grooming. Play behaviour is particularly important for young animals. Playing is not only rewarding in itself - it also serves an important function by helping animals learn the necessary social skills to function well in a herd as adults. Housed fattening cattle often lack the necessary space. The play opportunities of young dairy calves housed in small pens, sometimes individually, are also very restricted.

The potential to express normal behaviours varies according to the husbandry system. In well-managed extensive farming systems, cattle are able to express many highly motivated behaviours for much or part of the year. However, intensive high-production farming systems tend to limit or modify behavioural expression through commonly used husbandry practices, such as permanent housing, early separation of the cow and calf and mutilations.

Indoor environments tend to be monotonous and unstimulating. Well-managed outdoor environments allow for fuller behavioural expression through being complex, providing changing sensory stimuli, the choice to range and exposure to natural daylight, all things that permanently housed cattle never experience. Early separation of the cow and calf not only causes a period of distress (see F4), but also the mother loses the opportunity to express her highly motivated maternal behaviours, and the calf loses the benefit of her care and natural suckling. Mutilations such as dehorning (see F3) can also interfere with normal behavioural expression because horns serve various purposes, including communication with other cattle, thermal regulation, protection against predators, and comfort behaviours (horns allow cattle to scratch parts of their bodies that are otherwise difficult to reach).

However, other welfare issues may arise if the above-mentioned husbandry practices and other practices that restrict normal behaviours are not followed in current husbandry systems: the incidence of calf diseases can be higher if calves are left with their mothers; it may be much easier to meet the nutritional needs of some high-yielding dairy cows if they are permanently housed; injuries may be more likely if housed cattle are left with their horns. Welfare science can help assess the significance of these factors for cattle, but deciding whether such welfare compromises are acceptable, or whether practices should be refined or discontinued, ultimately remains an ethical judgement.

Cattle have been observed to develop abnormal behaviours in certain contexts, generally a sign that something they need is missing from their environment (see pp. 18–20). For example, intensively reared beef cattle that are fed a limited amount of long forage may perform tongue-rolling and excessive manipulation of inanimate objects, which can become stereotypical. Calves may exhibit ‘cross-sucking’, that is, suck the body parts of others in their group, when their normal suckling patterns are thwarted.

Welfare assessment protocols

Welfare Quality® has produced a general cattle welfare assessment protocol.[8] AssureWel has developed separate, specific protocols for dairy and beef cattle.[23] (see Chapter 3).

Advocacy examples

- Veterinary professionals who have gained the trust of cattle-farming clients are uniquely placed to encourage good stockmanship. This can be done by providing training to develop technical skills and discussion groups. It is helpful to avoid focusing solely on key performance indicators and spreadsheets as markers of herd performance, but also, spend time on individual animals and talk directly with stock workers about the importance of their work. Comprehensive SOPs should be part of effective health planning and include protocols for all husbandry on the farm, such as feeding regimes, treatment protocols for minor conditions, and tasks a farm operator is expected to perform. These should be readily accessible to all and translated as necessary. The veterinarian should ensure that everyone understands these SOPs and also promote effective staff communication.

- Veterinary professionals can also explore and introduce appropriate technological advances for better cattle management, such as the use of ear tags that monitor calf temperature, systems for remote monitoring of the calving cow, improved vaccines for disease prevention, and more reliable tests for carriers of infectious diseases.
- The longevity of the adult breeding herd is crucial for profitability, so veterinary professionals can encourage genetic selection that is not only focused on production traits but also on resistance to chronic diseases such as lameness and mastitis.
- Farm assurance schemes such as those promoted by supermarkets are powerful. They can make a real difference to farm animal welfare, for example encouraging health and welfare planning and demanding reduced lameness in dairy cows. However, they should focus on issues important to the cattle themselves, using evidence from animal welfare science studies. They must also identify non-compliance and take action accordingly. Veterinary professionals can play an important role in ensuring that farm assurance schemes achieve these objectives.
- EFSA has published detailed scientific opinions on the welfare of beef and dairy cattle (listed on p. 282), which can be valuable resources for veterinary professionals wanting to review the available evidence. The associated Plain Language Summaries (PLS) are especially helpful for time-strapped individuals. Reports summarise current research, identify priority areas for further study and make pragmatic recommendations for real-world change.

Sheep

9

Abstract

The core welfare concepts discussed in Part 1 are applied to farmed sheep. Common welfare hazards are listed. Common welfare outcomes are identified and structured according to the Five Freedoms framework. Practical welfare assessment protocols are referenced. Examples are provided of how veterinary professionals have advocated for sheep welfare or could be more involved.

Introduction

There are over one billion domestic sheep (*Ovis aries*) worldwide,[88] and they can be found on all continents. They are thought to be one of the first animals domesticated, with their ancestor being the wild mountain sheep, the Mouflon. There are now more than 850 breeds, selected by man and adapted to a huge range of environmental conditions, including arid desert conditions, pastures and mountainous regions. They are important to the rural economies of many countries, being used to produce meat, wool, milk and hides. Whilst sheep products comprise a very small proportion of the total global meat, milk and hide/fibre production, they remain economically and culturally very important. Sheep are adapted to and remain productive on land that could not support other forms of agriculture.

Most sheep are reared and managed in extensive, natural or semi-natural environments, in stark contrast to much conventional chicken and pig production, which takes place in highly artificial environments tightly controlled by man. It is likely that people perceive sheep as automatically having good welfare due to their leading a 'natural' life with considerable behavioural freedom. This partly explains the limited attention to and research into their welfare when compared with that of other commonly farmed species. However, this view is overly simplistic – welfare challenges are of a different type. There are also differences when comparing sheep in different production systems.

DOI: 10.1079/9781786391582.0009

This chapter focuses on the welfare of sheep reared for meat in temperate regions such as the UK, where meat is the main product. Wool remains important economically in other countries but is currently seen as a low-value by-product in the UK, with the cost of shearing barely covering the value of the fleece. Milk, cheese and wool production are important globally but are outwith the scope of this chapter.

Husbandry

Production systems vary considerably in terms of the environment, breeds used and the amount of human intervention.

Variations in the environment include the terrain and quality of pasture, whether sheep are outdoors year-round or sometimes housed, and whether supplementary feeding is routinely provided. Generally, breeds with traits best suited to the environment are chosen. The degree of human contact and intervention depends on the terrain, logistics and the number of sheep supervised by each person. This can range from just a few animals on a smallholding to thousands of sheep on a large commercial farm.

The UK has a stratified sheep production system with three tiers – hill, upland and lowland.

Hill farming in the UK is 'low input, low output'. Farming systems are very extensive with minimal human intervention, often limited to occasional inspections and the annual gathering. Sheep graze remote areas and are exposed to natural hazards year-round. They may be provided with supplementary feed in emergencies only. Breeds are typically small, adapted for harsher conditions, have good mothering abilities and carry a single lamb each pregnancy. Breeding is often completely natural, with a male ram introduced to the flock for a period of 6–8 weeks, timed to ensure a compact lambing period. Ewes are seasonally polyoestrous. In autumn, as natural daylight decreases, ewes enter their breeding period. They are pregnant during the winter and give birth to lambs in the spring. They enter anoestrus over the summer as the days get longer.

Upland and some lowland systems are still overall extensive, but there is usually more human intervention than in hill farming, with sheep fed and managed to optimise their productivity. For instance, there may be manipulation of reproduction with hormones used to advance breeding dates, and sheep will be fed and managed to promote multiple births. Sheep are often housed for time periods, usually mid to late winter, shortly before lambing. Breeds combine the mothering abilities of hill sheep with productivity traits of faster growth and larger litter sizes. They thrive in less harsh environments than hill sheep.

A minority of farming systems are intensive, characterised by near-continuous monitoring of the flock, full-time indoor management and tight control of breeding. Sometimes, invasive reproductive technologies, such as artificial insemination and multiple ovulation embryo transfer, are used to accelerate the genetic development and production potential of certain breeds.

Various routine husbandry procedures are performed in all farming systems: gathering and handling at intervals for vaccinations, body condition scoring, shearing, ultrasound scanning for pregnancy diagnosis, ear tagging for identification purposes, castration and tail docking of lambs.

Common welfare hazards

- Feeding regime.
- Provision of shelter and shade.
- Environmental conditions predisposing to foot damage.
- Parasite monitoring and control.
- Anthelmintic and antimicrobial resistance.
- Management and care of neonatal lambs.
- Mutilations (castration and tail docking in lambs).
- Handling quality and stockperson competence.
- Level of observation and record-keeping accuracy.
- Lack of welfare-focused research.

Welfare hazards are the resources (inputs) that can lead to poor welfare outcomes. The likelihood of poor welfare outcomes and their severity (i.e. the risks) can vary. Severity encompasses both duration and intensity. Common welfare outcomes are identified below, structured by the Five Freedoms framework.

Sheep welfare is determined by how well sheep can adapt, both physically and mentally, to their environment (see p. 8). Relevant descriptions of 'free-living' sheep are provided throughout to highlight the adjustments they must make while in human care. Here, 'free-living' is defined as sheep living freely in an environment with the resources for which they have evolved and making their own choices. However, such comparisons provide only part of the welfare picture; welfare science (see Chapter 2) is essential for a proper understanding of the welfare impact of different husbandry systems.

Key determinants of welfare are the knowledge, skill and motivations of stockpersons, as well as whether they have sufficient resources (see pp. 24–25). Where these factors are optimal, welfare will be as good as it can be within a given husbandry system. However, there are often persistent welfare issues that are challenging to resolve without altering the husbandry system itself.

Common welfare outcomes

F1. Freedom from hunger and thirst

The environments of extensive farming systems closely align with the natural environments for which sheep have evolved. To optimise welfare, husbandry needs to capitalise on the welfare benefits and mitigate the risks.

Sheep are ruminants that have evolved to spend most of their daylight hours grazing, during which they consume sufficient herbage to meet their nutritional requirements. Sheep in extensive farming systems graze in this manner for all or most of the year. Their innate behavioural tendencies, combined with learning, enable them to make the most of available nutrition. Access to food and water significantly affects flock movement patterns and home range sizes. In hill sheep, these can range from 25–50 hectares. Sheep have an excellent memory for the locations of their preferred food and water sources, even after just one visit to an area. They prefer moist grasslands over other herbage, and when given a choice, they consume the most palatable feeds first, which also tend to have the greatest nutritional value. Their cleft upper lip allows them to be highly selective and graze very close to the ground, while their continuously erupting teeth are sufficiently worn down by long hours of grinding herbage. They choose to drink daily when water is available. Many flocks become accustomed to and prefer to stay within certain areas, even if no physical barriers prevent them from roaming further, and they pass down local knowledge to their offspring.

Seasonal variation

Food availability and quality vary seasonally and annually with a changing climate, and grass growth almost halts in the winter. Breeding, pregnancy and lactation – all times of high metabolic demand – naturally occur at times of better herbage growth. However, there are times when availability does not meet demand, and this can lead to prolonged hunger and thirst, malnutrition and even production diseases.

Climatic extremes, such as summer droughts, winter snow and freezing conditions, can severely deplete feed and water supplies. While farmers can provide supplementary feeding at these times, this can be logistically difficult and expensive in very extensive environments. Provision may be easier when sheep are more accessible, such as when they are confined to lower pastures or can be housed.

Production demands

Farmed sheep, particularly those more intensively managed but even those extensively farmed, are under more pressure to produce than their wild counterparts. As a result, they tend to be kept in larger groups (flocks) and stocked more densely, increasing pressure on available resources. In some farming systems, sheep are often bred, fed and managed to be much more prolific than their wild counterparts, which raises their nutritional demands. Sheep can lose body condition and sometimes suffer from production diseases due to over-metabolising their body fat stores to meet the energy demand. For example, late gestation, pregnant ewes that are carrying multiple foetuses may be physically unable to fit enough food into their rumen.

Mitigations

Ideally, good stockmanship and management will prevent, detect early and promptly address certain common issues. Examples include the pain associated with lameness (Figure 9-1), which can prevent sheep from grazing sufficiently (see F3), and excessive wear of the incisor teeth ('broken mouth') in aged sheep that makes grazing difficult. 'Broken-mouthed' sheep require culling or moving to less harsh environments where they can cope for longer. In some areas, the soil may have insufficient mineral content, which in turn affects the quality of the herbage and leads to nutritional deficiency, even where sheep can graze to satiety.

When sheep are housed, a high level of stockmanship is essential to address differing challenges. For instance, it is important to ensure that all group members can access sufficient food. If sheep are overcrowded, subordinate animals may be pushed away from the feed and experience chronic hunger, despite the provision of sufficient feed for the flock overall.

Detecting nutrition-related problems in sheep can be challenging, particularly in extensively managed sheep. Sheep need to be gathered, handled and palpated to determine their body condition score, as their fleece hides their condition

Figure 9-1. Sheep commonly kneel to graze when they have lower limb pain.

from view. Each stockperson is often responsible for many sheep and must cover challenging terrain. However, body condition scoring usually provides valuable information on nutritional status. The production and welfare benefits often offset the extra work involved, and it should be part of key management stages.

F2. Freedom from discomfort

Extensively farmed sheep often have considerable behavioural freedom and choice to move to different areas that meet their comfort needs. They can typically tolerate a range of temperatures and climatic conditions provided they have access to shelter and shade to protect them from wind, rain, snow and sun, as well as access to dry-lying areas.

However, there will be challenging times, largely due to adverse weather conditions. Mitigations against discomfort in farming systems include ensuring that appropriate breeds are chosen for the environment and supported by good husbandry practices, such as bedding, supplementary feeding (Figure 9-2) and shearing as appropriate. Without these, sheep are likely to experience discomfort, such as heat and cold stress.

Most farmed breeds need shearing annually, according to the season. This contrasts with wild and primitive breeds of sheep, which naturally shed their fleece (moult) with the seasons and climate, helping them regulate their body temperature.

Well-managed housing can provide valuable protection from adverse weather at key times of the year and stages of life when sheep need extra support. However, good stockmanship is key. Overcrowding should be avoided, as it

Figure 9-2. Sheep given supplementary feed in winter.

predisposes to multiple problems. Examples include discomfort from poor air quality, heat stress and social tension among group members.

F3. Freedom from pain, injury and disease

Commercially farmed sheep require effective management to thrive. The natural environment does not always provide the essential resources they need, and breeding for productivity has increased their demands beyond those of wild sheep. Shortcomings in husbandry can predispose to non-infectious disease and accidental injury. These, in turn, can predispose to infectious diseases. Some highly infectious diseases can establish themselves even in fit and healthy animals. Good biosecurity can help prevent disease organisms from entering the farm and spreading within premises.

Some common injuries and diseases are outlined below to explore their welfare impact; further details on the actual conditions are available in numerous excellent clinical texts.

Care of neonatal lambs

Lamb morbidity and mortality are significant concerns, with nearly half of deaths occurring within three days of birth. Preventative measures, including ensuring lambs receive enough colostrum and bond properly with their mothers, are essential to mitigate this risk.

Farmed sheep may give birth to their lambs indoors or outdoors, presenting different risks and benefits. Outdoors, there are generally greater risks of death due to unnoticed dystocia, exposure to inclement weather and starvation of weaker lambs. Indoors, there are usually greater risks of infectious diseases due to the buildup of pathogens in the environment that occurs at high stocking densities. Lowland ewes tend to be more prone to dystocia than hill sheep because they are often fatter, with a narrower pelvis, and they carry multiple lambs.

Lameness

The cloven hooves of sheep are well adapted both for walking on soft plains and climbing rough, stony slopes. They grow continuously and are worn down by this constant activity. However, lameness is common in farmed sheep. It is used as a key welfare indicator, as it often causes long-term pain, reduces mobility and hinders normal behaviours such as grazing. Trauma, such as contact with sharp stones, predisposes the hooves to bacterial infections. Environmental factors, such as wet and muddy grounds and high stock density, further predispose. In 2011, the UK Farm Animal Welfare Committee recommended government targets for the industry to reduce lameness prevalence to 5% or less by March 2016 and to 2% or less by March 2021, but lameness levels of 10% or more are still common.[95] Changes in management practices can greatly reduce the risks.

Parasites

Ectoparasites are a common cause of skin damage in farmed sheep. Mite infestations cause severe itching, rubbing and licking, and if left untreated, can result in secondary infections and debilitation, which may ultimately be fatal. Sheep scab, caused by the mite *Psoroptes ovis*, is a particularly severe condition. Mites are common in cold climates and can spread easily in overcrowded indoor environments, especially among stressed animals with poor body condition scores and immunosuppression.

Cutaneous myiasis, also known as 'blowfly strike', is a serious condition that occurs when female flies lay eggs on soiled areas of sheep, commonly faeces stuck on wool at the hindquarters. When the maggots hatch, they devitalise and traumatise the skin as they feed. Preventative measures include good hygiene and fleece management, such as 'crutching' and shearing. Routine mutilations, such as tail docking (and mulesing in some countries), are often performed with the aim of reducing risks, but these mutilations produce another set of welfare harms (see p.149).

Endoparasites, like intestinal worms and fascioliasis, are also common problems, exacerbated by the rise of anthelmintic resistance. Holistic control strategies, tailored to individual farms, emphasise disrupting the life cycle of these parasites across the host and the pasture. Coccidiosis is a concern for sheep managed at high stocking densities, typically indoors.

Production diseases

'Production diseases' occur when the demands of production outstrip the body's ability to meet them (see F1). Examples include milk fever (hypocalcaemia) and 'twin-lamb' disease in peri-parturient ewes.

Trauma

Injuries can result from accidents, attacks and/or unskilful intervention by people. Accidents include becoming entangled in wire from poorly maintained fencing and running into obstacles when panicked. Attacks include predation of lambs by wild animals such as foxes and crows, and sheep of all ages being chased and bitten by pet dogs of irresponsible owners. Fighting is rare among sheep, except when introducing unfamiliar male rams or keeping groups of rams in confined spaces during the breeding season. Here, weaker animals can suffer severe injuries. Examples of unskilled interventions include overzealous and unhygienic interventions during parturition, rather than letting the ewe progress naturally, or failure to intervene early enough in cases of dystocia.

Mitigations

Preventative health and welfare planning, tailored to the individual farm and properly implemented, is key.

Early and accurate detection of health and welfare issues can be challenging. Firstly, sheep are a prey species that exhibit particularly subtle signs of pain. These often only become apparent once advanced. Only then do changes in activity, posture, feeding and interactions with other sheep become noticeable to those caring for them. Secondly, thorough surveillance of sheep can be difficult in many extensive farming systems. It can be challenging to get close enough to the animals to properly inspect them due to their fear of humans, the terrain and the sheer number of animals each stock person must observe. Thirdly, a lack of accurate recording can impede the identification of problems. Addressing supposed problems based on anecdotal information can divert improvement efforts away from where they are truly needed.

Mutilations

Mutilations, principally castration and tail docking, are common in many countries and production systems. For example, in the UK, male lambs are often castrated in the first weeks of life to prevent unwanted breeding, as some of these lambs could reach sexual maturity before they are ready for slaughter. Lambs of both sexes are commonly tail-docked to try and prevent 'fly strike' by minimising faecal contamination of the fleece at the hindquarters, thus making this area much less attractive to flies. Often, mutilations are performed without anaesthesia or analgesia. There has been considerable research into the welfare impact of the different methods used – rubber ring, surgical, and burdizzo/clamping – highlighting that some have more adverse effects on welfare than others. However, all have been found to cause considerable pain, and while mutilations are easier to perform in younger animals, they are likely to feel the pain as intensely as older ones (see p.30).

As mutilations can result in pain and distress, they should be avoided unless there is a clear and necessary welfare justification. Veterinary professionals can help ensure that decisions are made on a case-by-case basis as part of a well-considered health and welfare plan. When performed, every effort can be made to introduce anaesthesia and analgesia. The risks and benefits need to be properly weighed.

Veterinary professionals can inform the welfare debate, including raising awareness of the challenging context. Many vets understand the practical difficulties farmers face. For example, gathering hill ewes and their lambs and castrating the lambs with a rubber ring when they are less than one week old meets the legal age requirements for this procedure in the UK. However, these very young lambs often must cope with the pain of the procedure in a harsh, natural environment, and on balance, it may be more beneficial if the lambs were castrated at an older age. Also, there will be a spread of lamb ages within the sheep flock, so castrating and tail docking a wider age spread of lambs within a single gathering may, on balance, be better for welfare and production than subjecting the flock to the stress and disruption of repeated gatherings in order to ensure all mutilations are done within the first week of life. Such factors need to be properly explained and inform the welfare and legal debates. There is a useful open-access paper by Gascoigne *et al.* (2021)[56] that applies the ethical concept of the 3Rs (see Chapter 4) to castrating and tail docking of sheep.

Mulesing is a very painful mutilation, illegal in the UK, but performed to prevent flystrike in some breeds and countries, particularly Merino sheep in Australia. It involves removing excess skin folds from the sheep's perianal region using wool trimming shears, without the use of anaesthesia or analgesia. Scar tissue forms as a result, which is smoother, less likely to get dirty and, therefore, less attractive to the flies.

There are horned varieties of sheep, and, unlike many dairy cattle and goats, they are not routinely dehorned. Sheep may use their horns to defend themselves, provide balance and regulate heat loss. The size of horns may also reflect social ranking.

F4. Freedom from fear and distress.

Unpleasant feelings such as fear and anxiety often accompany pain, injury and disease. They can also occur independently. Momentary fear and alarm are a normal part of everyday life, but heightened and/or prolonged fear can be a welfare issue.

Sheep are a prey species, easily fearful. They have evolved with certain senses and innate behavioural tendencies to protect themselves from threats. For example, sheep possess wide-angle, long-distance vision that allows them to readily detect movement. They have evolved to live in groups and have a strong, innate desire to be with other sheep. Flocking behaviours, where they follow and move closely with others in the group while walking, running, grazing and resting, are very important to them. Regardless of habitat, sheep tend to locate themselves near terrain that they can escape to when they feel threatened. At vulnerable times, like when resting, they often prefer to be on higher ground (hilltops and ridges) that provides better vantage points.

In many well-managed farming systems, sheep can continue to express these behaviours, which is likely to make them feel more secure. However, at times, routine husbandry places them in environments and situations they find threatening.

Human–sheep interactions

Sheep intermittently need to be moved and handled for routine husbandry procedures such as shearing, crutching, foot bathing, vaccination, dipping, weighing, sorting, loading, scanning and body condition scoring – all of which can be aversive to them but also yield important health benefits. During such procedures, the flock often must be held stationary in a confined space, something not experienced by sheep in a natural environment, which would either be caught by a predator and killed fairly rapidly or escape. They can perceive their human caretakers as potential predators.

Fear behaviours are also used routinely as a handling tool in farming. When sheep feel threatened, they become agitated and flee as a flock, provided they have the space to run. Farming routinely makes use of this, strategically using humans, vehicles and dogs to create and control the direction of movement. For many extensive systems, there are few other practical strategies.

Handling quality

The level of stockmanship is a key determinant of how well the sheep cope (Figure 9-3). Handlers must stay mindful of the fear sheep are experiencing; this can be easily overlooked due to the subtlety of their behavioural signs. Often, the sheep remain quiet and very still, bunching tightly together, which can be misinterpreted as calm rather than fear. They only tend to vocalise and show evident agitation when very distressed, such as when separated from the rest of the group or when ewes are separated from their lambs.

Handlers should use best practices to prevent stress from escalating to distress. For example, welfare can be improved by avoiding vigorous movement by dogs and humans, minimising noise and preventing dog bites. Rough handling is aversive and can also easily cause pain and injury in such a small animal that can be easily overpowered manually. Sheep should never be grabbed and lifted by the fleece. Note that sheep have a good ability to learn and to remember places, events and people. They are capable of associating places with negative

Figure 9-3. Trained sheepdogs are often invaluable for gathering sheep in remote terrain but must be used carefully to ensure the animals do not become overly fearful and distressed.

experiences for up to a year and remembering specific handlers for at least two years. Whilst they can habituate to frequent gentle human handling, there is little practical opportunity for this in extensive sheep farming.

An effective handling system, with features that support better welfare, can be instigated with forward planning. For example, it can exploit normal sheep behaviour by ensuring they can see and follow each other at all times, and it should provide enough room for all animals. Sheep can be allowed time to quietly settle and familiarise themselves with their new surroundings.

Social stress

It is beneficial that farmed sheep typically remain in stable social groups, allowing them to become familiar with other individuals and form their normal social hierarchies. However, if the farming system places them in situations where they must compete excessively for important resources, such as when confined and overcrowded indoors, social interactions can become unfriendly and create considerable social stress. Stockpersons need to be aware of and plan around these higher-risk times.

In sheep, unfriendly interactions may be overlooked, unless severe, as they generally present as subtle eye contact and subtle movements, such as resting their chin on another sheep's back or nudging with their front leg, rather than overt aggression. Exceptions are the obvious aggressive behaviours displayed by ewes protecting their lambs and rams clashing, particularly during the breeding season.

Social isolation

It is usually possible to avoid extreme fear and distress caused by social isolation by planning ahead and having management systems that keep sheep together. Restraint-fostering practices, where sheep are separated and held by the neck in stocks to induce acceptance of lambs, can be replaced with more welfare-friendly methods.

F5. Freedom to express normal behaviour

Domestic sheep are known to retain many behaviours of their wild ancestor, and basic ethograms exist. However, our understanding of sheep behaviour is less advanced than for other commonly farmed species. For instance, their learning abilities and memory have only recently been acknowledged, and there has been little research into their motivations and preferences.

Behavioural needs

It is beneficial for the welfare of farmed sheep that they typically retain considerable choice and behavioural freedom in a habitat for which they are well adapted, despite some restrictions on their movement, social interactions and diet. Some important normal behaviours have already been identified under the other Freedoms. For example, they can choose their movement patterns and locations when grazing outdoors, and have the freedom to select different plant types and graze continuously (see F1). When housed, their choices are fewer, but if housing is well managed, they will instead benefit from the protections housing provides, such as shelter from inclement weather and additional feed provision (see F1, F2). Farmed sheep are generally able to maintain social behaviours important for their well-being, living and functioning as relatively stable groups, and they are rarely isolated (see F4).

Behavioural 'wants'

Welfare involves providing as many positive experiences for the sheep as possible, rather than just avoiding aversive ones. Below are some additional ideas on how normal behavioural expression may enhance welfare, but these are as yet little researched.

Sheep are likely to have a fundamental desire to forage and explore, as rewarding activities in themselves, beyond meeting the purposes of obtaining enough food. Welfare research could provide insights into 'enrichment' for housed sheep, optimising pen layouts and how feed is provided. Sometimes, small, evidence-based tweaks can make a large difference to animals' experiences.

Specific social relationships appear to matter to sheep, beyond their basic need to live in a group, so it is beneficial that they can often sustain these in farming systems. For example, 'free-living' sheep usually choose to associate with preferred companions and to select their subgroup for activities such as grazing. Stronger bonds tend to develop between particular ewes, based on family ties or past rearing experiences, and ewes seem motivated to maintain these associations. They recognise one another through a combination of sight, sound and smell, releasing distinct scents from glands located in their feet, inguinal area and face. Social behaviours among sheep are often subtle, so these 'friendships' can easily go unnoticed, even by those who work closely with them. Friendly behaviours typically manifest as sheep remaining near specific individuals while grazing or lying down, rather than through overt actions such as mutual licking and grooming. An exception to this is when ewes lick their lambs shortly after birth.

Ewes and lambs form strong attachments, with wild ewes rarely separating from their lambs until they are at least 6 months old, and certainly not in life-threatening situations. In contrast, farmed lambs are usually weaned much earlier, around 12–14 weeks of age. However, this still provides a significant period during which the ewe and lamb can benefit from engaging in normal mothering and suckling behaviours.

Some other naturally changing social structures are partially replicated in the farming system, such as the formation of separate male and female groups outside of the breeding season.

Abnormal behaviour

It is positive that little 'abnormal' behaviour has been noted in farmed sheep, though this area needs further investigation. Again, it may be that abnormal behaviours in sheep are subtle and easily overlooked. When abnormal behaviours such as stereotypies do occur, they tend to be observed in sheep under very restricted conditions, for instance in certain experimental settings. If farmed sheep are housed indoors at high stocking densities with limited access to feed and fibre, they may engage in behaviours such as nibbling, biting and chewing the wool of other sheep.

Welfare assessment protocols

AssureWel and AWIN (see Chapter 3) have produced welfare assessment protocols for sheep.[22,23]

Advocacy examples

- Accurate identification of health and welfare problems can often be challenging due to inadequate record-keeping and overreliance on anecdotal evidence. There is substantial room for improvement, whether records are electronically linked to individual sheep or consist of written batch records for flocks. Veterinary professionals can support sheep farmers in this, ensuring that efforts and resources target real problems.
- Veterinary professionals can raise farmer awareness of programmes such as the Food Animal Initiative (FAI) 'Lameness – Five-Point Plan' for the UK sheep industry.[95] This initiative challenges the perception of lameness being 'normal' in sheep production. Results from FAI farms showed improvements in both welfare and production due to reduced levels of lameness, and led to its adoption by groups like the Sheep Lameness Stakeholder Group and the RSPCA Sheep Welfare Standards.[96]
- Veterinary professionals can raise awareness of the dearth of sheep welfare research and standards, for example, by building on the EFSA scientific report on sheep welfare (2014).[97] Veterinary professionals can assist the research community to understand the husbandry and socioeconomic context of sheep farming, as this determines the key welfare science questions from a pragmatic perspective.
- There is currently limited welfare labelling for sheep products, but useful developments are underway that are worth staying informed about. In the UK, RSPCA Assured, which implements the RSPCA Sheep Welfare Standards[96] is developing sheep welfare standards that now incorporate AssureWel protocols and the Five-Point Lameness Program.[96] Their Sheep Standards Technical Advisory Group comprises credible stakeholders, and

meeting summaries are available online. Discussions cover a diverse range of topics, including outcome-based welfare assessments, grazing management, parasite control, stockperson training, the impacts of dog worrying, and the effects of Brexit on sheep welfare.

- Veterinary associations can lobby governing bodies involved in trade to uphold UK sheep welfare standards and to prevent them from being undermined by imports with less stringent welfare standards. A notable example is the BVA's response to the International Trade Committee's inquiry regarding UK trade negotiations with Australia.[98]
- Veterinary associations should invest in preserving corporate memory to ensure veterinary professionals are aware of past initiatives. This enables them to build on previous experiences, avoid repeating similar ideas without reflection, and make more informed, productive progress. Modern digital technologies should facilitate the creation of searchable digital archives of past initiatives.

10 Dogs

Abstract
The core welfare concepts discussed in Part 1 are applied to dogs. Common welfare hazards are listed. Common welfare outcomes are identified and structured according to the Five Freedoms framework. Practical welfare assessment protocols are referenced. Examples are provided of how veterinary professionals have advocated for dog welfare or could be more involved.

Introduction

Domestic dogs (*Canis lupus familiaris*) are descended from the grey wolf (*Canis lupus*). The dog has been domesticated for longer than any other mammal, around 12,000 years, and has lived in particularly close association with humans during this time. Domestication has altered cognition in dogs to the extent that using wolf behaviour to understand dog behaviour is unhelpful, even counterproductive.

Today, dogs are found wherever humans live, interacting with and depending on people to varying degrees. They may be completely under human control, or free-roaming dogs that loosely associate with a particular community, otherwise known as 'street dogs', or feral dogs that coexist alongside human settlements. The exact number of dogs worldwide is not accurately known. Global estimates are of around a billion dogs.[99] About half are believed to be stray or feral, while the other half are presumed to be owned companion animals.[100] These numbers are probably underestimates, as dogs are not registered in many countries.

Humans use dogs in a variety of ways, and attitudes towards dogs vary amongst cultures and societies. They may be companion animals (pets) whose primary purpose is to provide humans with social interaction and enjoyment. They may be primarily working dogs, for example, those used in hunting, herding or guarding, sniffer dogs for locating drugs and landmines, assistance dogs such as guide dogs, and emotional support animals. They may be used in sports, such as racing greyhounds, or other forms of human entertainment,

DOI: 10.1079/9781786391582.0010

including dog showing or even dog fighting. In some countries, they may produce meat for human consumption.

They may be viewed affectionately as treasured companions; dogs often fill social needs in humans, creating strong bonds between humans and dogs that are comparable to or even exceed those between people. They may be disliked for the nuisance they cause or feared due to the risk of zoonotic infectious diseases, particularly rabies in areas where the disease is endemic.

Dogs are a massively diverse species. Humans have created over 400 dog breeds worldwide, with such differing shapes, sizes and behavioural tendencies that it can be challenging to describe the typical dog. However, certain traits are common to all dogs. They are social animals and require the company of other dogs or humans (Figure 10-1). Activities they often choose to engage in are socialising with other dogs or humans, exercising, playing, exploring, resting in a comfortable place and eating. They share a common system of communication and an acute sense of smell and hearing. Scent is particularly important in canine communication, and dogs prioritise sniffing each other when they meet. Urine marking is part of their territorial behaviour.

Domestic dogs are unusual in that they appear to find human company particularly rewarding, perhaps even more so than that of other dogs. They are uniquely placed to live alongside humans in mutually beneficial relationships, which can provide each with the social contact they need for a good quality of life. However, there are numerous potential welfare issues, many of which relate to breed standards, the impact of modern human lifestyles on dogs and people misunderstanding dog behaviour.

Figure 10-1. A group of dogs out for a walk together, demonstrating the diversity of breeds.

Research into the welfare of domestic dogs is growing, but there is still far less than that of farmed animals. Little is known about how dogs would live if left to their own devices outside human control. Even so, an assessment of dog welfare can be made using key welfare concepts.

This chapter focuses on the welfare of dogs in two broad contexts: owned companion (pet) dogs and free-roaming 'street dogs'.

Husbandry

Companion dogs

In many countries, it is normal for dogs to live in households, closely alongside and owned by people as companion animals, who often tightly control all aspects of their lives. Many veterinary professionals are familiar with this model of ownership.

Street dogs

In Western countries, dogs tend to be categorised as either 'companion animal' ('pet') or 'stray' depending on their ownership status. This classification is overly simplistic and inappropriate for many other countries, such as India. 'Street dogs' are dogs with a high degree of autonomy, often not owned by or considered the responsibility of any one person. People have varying relationships with these dogs, ranging from avoidance to coexistence, tolerance and symbiosis. Attitudes may differ between city areas, between streets and even within the same street, and this directly affects dog ecology. Although street dogs are free-roaming, they are not reproductively capable of maintaining their population density unless food and harbourage are provided by humans. Provision may be intentional or non-intentional in the form of rubbish dumps and empty buildings. Certainly, human care for pups greatly reduces their mortality in the early months of life. Attitudes are, in part, determined by the risks to human welfare. The spread of zoonoses is a problem where high densities of dogs and humans live in close association. Veterinary professionals may become involved in projects aimed at controlling dog populations humanely, a strategy that can benefit both dog and human welfare.

Common welfare hazards

- Feeding regime.
- Breed standards.
- Population control approaches.
- Level and duration of confinement.
- Level of socialization.
- Social isolation.

- Training methods.
- Level of veterinary care.
- Mutilations.
- Owner knowledge of dog behaviour.

Welfare hazards are the resources (inputs) that can lead to poor welfare outcomes. The likelihood of poor welfare outcomes and their severity (i.e. the risks) can vary. Severity encompasses both duration and intensity. Common welfare outcomes are identified below, structured by the Five Freedoms framework.

Dog welfare is determined by how well dogs can adapt, both physically and mentally, to their environment (see p. 8). Relevant descriptions of 'free-living' dogs are provided throughout to highlight the adjustments they must make while in human care. Here, 'free-living' is defined as dogs living freely in an environment with the resources for which they have evolved and making their own choices. However, such comparisons provide only part of the welfare picture; welfare science (see Chapter 2) is essential for a proper understanding of the welfare impact of different husbandry systems.

Key determinants of welfare are the knowledge, skill and motivations of people caring for dogs, as well as whether they have sufficient resources (see pp. 24–25). Where these factors are optimal, welfare will be as good as it can be within a given husbandry system. However, there are often persistent welfare issues that are challenging to resolve without altering the husbandry system itself.

Common welfare outcomes

F1. Freedom from hunger and thirst

Unlike cats, dogs are omnivores that can thrive on meat-free diets.

Companion dogs

Companion dogs will suffer from prolonged hunger and thirst if underfed and malnourished. However, obesity appears to be a more pressing welfare issue, resulting from overfeeding and inappropriate food types, combined with a lack of exercise. This perhaps mirrors trends in many human populations. Obesity impacts the other Freedoms. For instance, it predisposes to or exacerbates health problems such as arthritis, diabetes mellitus, cardiovascular problems and skin intertrigo. It can limit the dog's activity and normal behaviours that they would otherwise enjoy.

Street dogs

Hunger and thirst levels in street dogs vary based on the amount, quality and frequency of food and water people provide, intentionally or unintentionally as poorly secured rubbish. When resources become scarce, competition from other dogs can lead to hunger and malnutrition, especially in animals with higher nutritional needs, such as pregnant and lactating female dogs. Where stray or street dogs are captured, rehomed and housed in shelters, they may still experience hunger and thirst due to factors such as overcrowding. Levels of care at these facilities vary significantly.

F2. Freedom from discomfort

Companion dogs

Companion dogs are often considered part of the family, and a comfortable, dry, warm resting area is usually provided. However, other aspects of comfort may be overlooked by humans as dogs live in a very different sensory world. They have a far superior sense of smell and hearing and can detect and be affected by scents and sounds, both inside and outside the home, to which humans are oblivious. Examples are scents from cleaning agents or very high-frequency noises inaudible to humans.

Street dogs

The comfort of street dogs depends on the availability of shelter, shade and dry resting areas, and whether these are sufficient to protect them from the climate. Again, access may be limited by competition from other dogs when resources are scarce. After being caught and placed in kennels and rehoming centres, the welfare of dogs depends on the facility standards, which can vary greatly. Poor environmental conditions can cause discomfort, and once captive, the dogs are unable to move freely to seek better conditions.

F3. Freedom from pain, injury and disease

Breed standards

Many companion dogs are of a specific breed. Humans initially selected dogs for tasks that required good fitness, such as herding, guarding and hunting. However, since the mid-1800s, breeding solely for appearance has become common in many countries. Most breed standards, which are guidelines for the appearance of different dog breeds, focus solely on appearance without considering the health and welfare of the dogs. Many countries have a central body, such as the Kennel Club in the UK, that governs these standards. Breeders strive to meet these standards and can receive prizes in shows as a reward.

Breed standards often dictate harmful extremes in dog anatomy. These extremes can cause pain and discomfort. For example, the excessive skin folds of a Sharpei frequently cause chronic pain, the narrow airways of brachycephalic breeds such as pugs make breathing and exercising difficult, and the narrow pelvis of bulldogs means they cannot give birth without a caesarean section. Breed gene pools are small. This predisposes certain breeds to harmful inherited disorders, such as hip dysplasia in Labrador retrievers and mitral valve heart disease in Cavalier King Charles. Many dog breeds could not be sustained without veterinary interventions. Treating individual dogs presented in the clinic without also addressing the root causes could perpetuate breed-related welfare problems.

Indiscriminate breeding

Indiscriminate breeding and selling of companion dogs leads to health and welfare problems associated with overproduction, failure to use canine health check schemes, inadequate facilities, inadequate socialisation of puppies, general lack of health checks and inadequate worming and vaccination.

Level of veterinary care

Dogs are likely to experience pain, injury and disease when their owners fail to access veterinary care due to cost, lack of awareness, or lack of interest. Preventative treatments such as flea control, deworming and vaccinations, along with timely veterinary care when pets are sick or injured, are important. Veterinary practice has advanced greatly in recent decades, offering a wider range of diagnostic tests and treatment options. This allows companion dogs to recover from or live happily with conditions that at one time would have caused suffering or ended their lives, for example, diabetes. However, an owner's desire to do everything possible for their dog can lead to 'overtreatment', that is, clinical interventions that go beyond those that support and protect the dog's overall quality of life. Dogs are also living longer. As a result, there is a growing need for information on managing the quality of life of geriatric animals, such as chronic pain associated with conditions such as osteoarthritis. 'Quality of life' assessment protocols developed in research are beginning to be used in veterinary practice.

Mutilations

Companion dogs undergo various mutilations, which are defined as procedures that interfere with the sensitive tissues or bone structure, performed for purposes other than medical treatment. Some types of mutilation may be justified for overall welfare benefits, while others cannot. The extent of pain and distress depends on the type and method of mutilation. Microchipping involves inserting a tiny identification device under a dog's skin using a large-bore needle. It causes minor, momentary discomfort, which is offset by major benefits, such as reuniting lost dogs with owners and ensuring owners remain accountable for their dogs' whereabouts and condition. This procedure is a legal requirement in many countries. Elective desexing (ovariohysterectomy or castration) is a common mutilation in many countries with well-developed veterinary services. It can provide welfare benefits, such as preventing welfare harms associated with the birth of unwanted puppies. Not all owners can manage their dogs' behaviours when in oestrus. However, desexing, particularly ovariohysterectomy, is an involved operation, not without risks. A thorough discussion with the owner about the procedure's pros and cons is preferable to insisting on desexing every dog in every situation. For example, some owners have considerable control over the whereabouts and actions of their dogs, which alters the welfare risks.

Other mutilations, such as tail docking, ear cropping and debarking, are very difficult to justify. They all cause pain and distress and do not benefit the dog. Tail docking involves removing part or all of the tail using sharp scissors or a blade when the puppy is a few days old, usually without any anaesthetic or analgesia. Historically, tails were docked for various reasons, from a presumption that it would help control rabies to attempts to prevent injuries to working dogs' tails. Nowadays, most tail docking is done for cosmetic reasons, even in working dog breeds. Dogs with docked tails are only marginally less likely to suffer tail injuries than dogs with entire tails, and effective modern treatments for tail injuries are readily available. Docking causes severe, acute pain and distress. Neonates feel as much pain as older animals. It is unclear if chronic pain or hypersensitivity develops at the healed tail stumps. Ear cropping involves removing part of the ear pinnae so that the remainder stands erect, and is done for cosmetic reasons. Debarking (devocalization) involves removing tissue from the vocal cords to reduce the volume and harshness of the bark sound, and is done for the owner's convenience (ear cropping and debarking are prohibited in some countries, including the UK).

Street dogs

Street dogs tend not to experience breed-related problems, as they are not subject to the controlled breeding by humans required to maintain breed standards. Without this, dogs naturally tend towards the typical street dog phenotype, which supports fitness and function. However, street dogs are more exposed to other risks than responsibly owned dogs. Examples are high rates of spread of infectious diseases and parasites, accidents such as road traffic accidents and lack of access to veterinary preventatives, treatments and euthanasia.

Due to a lack of resources, understanding or appreciation of welfare, government authorities often attempt to control dog populations by catching and killing them in an inhumane manner or poisoning them *in situ*. Simply killing dogs is ineffective for population control. Even maximal catching and killing rates have little impact, as other dogs quickly migrate into the area, filling the ecological niche. Internationally recognised organisations such as the World

Organisation for Animal Health (WOAH), the World Health Organisation (WHO) and the Food and Agriculture Organisation of the United Nations (FAO), along with various veterinary associations, recommend habitat control, responsible ownership and surgical neutering as humane methods to manage dog populations and control zoonotic diseases, including rabies. Human and dog welfare should improve in tandem.[101]

'Trap-neuter-release' programmes

Veterinary professionals are essential for properly addressing the above issues and are often involved in programmes initiated by animal charities in various cities globally. They may work locally or abroad, needing to adapt their skills to fit a very different context.

In trap-neuter-release/return (TNR) programmes, sometimes called Animal Birth Control (ABC) programmes, street dogs are captured, neutered, vaccinated and released back into the same area to produce a smaller, stable, healthier dog population. Surgical neutering is currently the only completely effective method, but other infertility methods may be developed in the future. If ABC programmes are well managed and acceptable to the local community, they can benefit both dogs and humans. If poorly executed, they can cause considerable welfare harm. There are many factors to consider. These include the skills, knowledge and attitudes of programme staff, local resources and handling and catching techniques. A different surgical approach may be advocated than that used in owned dogs. For example, spaying animals using a flank approach to enter the abdomen may lower the risk of herniation in situations where post-operative care is limited and suture material types are restricted. Surgeons need to be competent. For instance, veterinary students volunteering in these projects must be properly supervised by those competent, ensuring good standards. Ethically, they need to keep their motives in check and not view street dogs as unowned, 'practice' animals when being given the opportunity to perform more procedures with fewer controls than in their own country.

There are welfare and ethical issues associated with prematurely releasing dogs from veterinary care, around 3–5 days post-operatively, before wounds have completely healed. Street dogs are active and unsupervised. The dog's ability to cope on the street may be affected by complications such as wound breakdown, seromas, infections and pain. Balancing the need to release dogs shortly after surgery – both to maintain a high surgical throughput and to reduce the stress of being kenneled – with the requirement to ensure that the dogs are fit to cope and that post-operative complications are minimised, can be challenging.

F4. Freedom from fear and distress

Fear and distress are often associated with pain, injury and disease. They also occur independently.

Owner knowledge of dog behaviour

Humans frequently misunderstand dog behaviours. In many countries, there are no recognised standards for, or regulation of, dog trainers, which means some are providing advice harmful to dogs. Organisations such as the Animal Behaviour Training Council in the UK are seeking to address this. For example, outdated comparisons between dogs and wolves see dogs constantly striving to achieve status, viewing human cohabitants as the 'pack' they need to dominate. This can lead to ineffectual physical punishments that create fear and anxiety, such as 'shaking down' to put the dog in its place. Many 'behavioural problems' in dogs stem from fear rather than a desire to dominate. For instance, biting is often a last resort response to fear once the dog has tried all other options. People often perceive their dog to have suddenly bitten, having missed subtle, preceding signs of fear and anxiety.

The use of punishment to train dogs often causes fear and distress. Training aids that use punishment include electric collars, which apply an electric current to the skin. They can be remotely activated by a handler or activated by a stimulus from the dog wearing it, such as barking sounds. In countries where their use is legal, there are few controls on who can use them and how. Fear and distress are exacerbated by misuse, including excessive current and improper timing of application. They are also unnecessary, as dogs can be humanely and enjoyably trained using reward, which is also far more effective than punishment (see pp. 20-24).

Level of socialisation

Puppies have a 'critical' (sensitive) period between 3 and 11 weeks of age. Socialisation – positive exposure to people, animals and the environment – is essential during this time. Without it, they may develop long-lasting fears and anxieties in everyday situations. These experiences can shape their personality for the rest of their lives.

Street dogs are automatically exposed to many sights and sounds from birth. However, companion dogs will not be unless people actively expose them. Dogs from impoverished environments, such as so-called 'puppy farms' that breed dogs on a large commercial scale to produce puppies for sale, are at particular risk of a lack of socialisation. This crucial aspect of their development may be overlooked entirely, or there may simply be too few people available to ensure appropriate human–animal interaction.

Owners commonly report that their dogs show extreme behavioural responses to certain noises, such as fireworks, thunderstorms and gunshots. Unfortunately, these tend to be inadequately treated as they occur intermittently, and can easily become low priority for owners between episodes.

Separation anxiety

Dogs are social animals, so when left alone by their owners, they can experience separation anxiety. They may show signs such as toileting in the house, barking and destroying objects, especially near the door through which their owner left. Even the presence of another dog may not fully alleviate their distress. Many cases of separation anxiety in dogs may go unnoticed and unresolved, particularly when the dog shows more subtle signs of distress or when the behaviour fails to trouble the owner.

Relinquishment to rehoming/rescue shelters

Unresolved behavioural problems often lead to dogs being given up to rehoming or rescue shelters. Many dogs in these shelters already have psychological issues and are then further burdened with the stressors of confinement and isolation, which worsen any existing fear and distress. This can lead to these dogs being euthanised. Many problems might be prevented by proper socialisation at a young age.

Veterinary clinic visits

Dogs often experience fear and distress in alien environments and environments they have learned to associate with unpleasant experiences, such as visits to veterinary practices. To mitigate this, veterinary practices can strive to make their premises and approach more dog-friendly. Examples include using sound-absorbing materials in the kennel area, adjusting lighting, providing retreat areas with familiar-scented objects, and having all staff trained in dog behaviour and welfare. 'Fear-free veterinary practice', which incorporates these approaches and emphasises the dog having the best possible experience during a clinic visit, has gained popularity in some places.

Street dogs

Street dogs can experience fear and distress when suffering pain, injury or disease, and in response to other emotional stressors. Whether the fear is manageable or tips over into distress depends on its duration and severity, as well as whether the dog can take actions to help itself and feel some sense of control.

F5. Freedom to express normal behaviour

Companion dogs

Humans often closely control the lives of companion dogs, deciding when and where the dog sleeps, exercises, eats, defecates and urinates, with whom the dog interacts and the nature of any interaction, leaving the dogs with limited choices. Dogs can struggle to cope with busy, modern human lifestyles. They may experience boredom, frustration and anxiety, expressed as various behaviours. Usually, 'behavioural problems' are normal dog behaviours that cause owners a problem. Any attempts to correct the behaviour may exacerbate it if owners misunderstand dog motivations and how they learn.

Owners can inadvertently teach the dog behaviours they do not want. An example would be dogs that receive attention they find rewarding for excessive barking, jumping and overexcitement. Another example is the dog that initially fails to return to the owner when called. When the dog does return, the owner may be so frustrated that they punish the dog for just performing the very behaviour they want.

Many breed standards limit the dogs' ability to express normal behaviours. Their extreme conformation may prevent them from communicating effectively with other dogs as they cannot adopt the correct body language, thus reducing social harmony. For example, dogs with excessive skin folds cannot adopt different facial expressions. Those with very short limbs find it difficult to perform the play bow, and those with curled, rigid or docked tails cannot effectively use tail position to signal their mood and intent. Dogs selectively bred for specific roles are known to have differing motivations/behavioural tendencies; for example, the border collie is strongly motivated to herd and chase. However, it is unclear how specific breeds feel if these behaviours are thwarted.

Specific mutilations also limit the ability to express normal behaviours and are not offset by any welfare advantages. Removing the tail (tail docking) reduces the dog's ability to communicate. Debarking does not address the underlying issues that cause the dog to bark, such as stress when left alone; it simply removes the inconvenience of the noise for humans.

Age can also limit behavioural expression. For example, elderly dogs can reach the point where their mobility is limited by osteoarthritis, and their sight and hearing are deteriorating. However, with creative thinking and action, they can still be provided with a life worth living. A practical example is carrying or pushing the dog in a trolley to a familiar park it can no longer walk to, which allows it to still enjoy exploring using its sense of smell.

Street dogs

Free-living street dogs generally have considerable behavioural freedom and autonomy. However, some behavioural unknowns related to trap-neuter-release programmes could be addressed by further research. For example, one question is whether releasing dogs into the area of the city where they were caught, but not necessarily on the same street, is adequate. Another question is how neutering affects dog social interactions. Little is known about the bonds formed between street dogs; even less is known about the effect of neutering on these and the welfare impact.

Welfare assessment protocols

Companion dogs in the home environment

Tools are being developed to assist veterinary professionals and owners in holistically assessing quality of life (QoL). Many are still to be validated. QoL tools generally revolve around owner-completed questionnaires and share many features of some of the better-known chronic pain scoring systems that have been available for longer. There are various barriers to implementation in veterinary practice, including a lack of awareness, time constraints and economic considerations.[102]

The PDSA's PetWise programme in the UK has provided veterinary professionals with training in general welfare/quality of life assessments for dogs, cats and rabbits. A simple 'traffic light' system was used to highlight key strengths and to identify priority areas for improvements, and to develop an action plan in conjunction with the owner.[103]

There are assessment tools that focus on measuring acute and chronic pain and welfare in dogs affected by specific health conditions, as set out in the WSAVA Pain Guidelines.[44] Ongoing initiatives, such as VetMetrica,[45] are making these tools more user-friendly and readily available on electronic devices. They are also using data to benchmark scores and help users interpret the scores.

Dogs in shelters and rehoming centres

- The Dog Welfare Assessment App, developed by the Jeanne Marchig International Centre for Animal Welfare (JMICAW), is useful for assessing the welfare of dogs in trap-neuter-release programmes.[104]
- The SHELTER QUALITY Welfare Assessment Protocol for dogs housed in rescue and rehoming centres, developed in 2016, was inspired by the Welfare Quality approach.[105]

Advocacy examples

Dog breeding

- Consultations and investigations have laid the groundwork for addressing welfare concerns in pedigree dog breeding in the UK. Key reports from 2009 and 2010 include those instigated by the RSPCA,[106] the Associate Parliamentary Group for Animal Welfare[107] and Professor Patrick Bateson.[108]

  Veterinary professionals and other stakeholders worked together to take practical action based on the evidence provided by the above reports, forming independent bodies and information hubs, such as the Advisory Council on Dog Breeding Welfare, the Dog Breeding Reform Group,[109] and UFAW's webpage 'Genetic Welfare Problems of Companion Animals'.[110]
- In 2012, the RSPCA and the BVA Animal Welfare Foundation collaborated to create and launch the Puppy Contract, a free tool designed to encourage responsible puppy breeding and purchasing.[111]
- The BVA Brachycephalic Working Group discussed the 2018 film 'Patrick', which featured a pug dog, with the multimedia conglomerate Disney.

As a result, Disney agreed, among other things, to include a welfare message in the credits about pugs' health issues and not to produce 'Patrick' pug merchandise. The BVA also offered the film industry future support to ensure that animal welfare implications are considered early at the design stage.
- The BVA now provides a template letter for veterinary professionals to easily express their concerns to advertisers and media outlets that inadvertently promote brachycephalic breeds in their materials.[112]
- In 2014, the UK parliament discussed puppy farms following a successful petition. In 2020, 'Lucy's Law'[113] was enacted, mandating the licensing of breeding businesses. Citizens must either buy puppies and kittens directly from breeders or adopt them. When purchasing from a breeder, they must see the puppies with their mothers in their birthplace.

Understanding dog behaviour

- In 2024, several international veterinary associations, including the FVE, WSAVA, FEEVA and FECAVA, released a joint paper on animal behaviour and training. Key themes include protecting welfare by ensuring proper socialisation and habituation of young animals, using positive reinforcement rather than aversive training methods like electric pulse collars, and encouraging the accreditation of animal behaviourists. They also emphasise that animals need to express certain natural behaviours.[114]
- Veterinary professionals can support other stakeholders in raising welfare standards. For example, the UK Animal Behaviour and Training Council is working to further regulate and accredit its members, which could impact many companion animals in the UK.
- Veterinary professionals with behavioural expertise have found ways to engage and educate large sections of society on dog welfare issues. An example is the television programme 'Dogs: Their Secret Lives', which aired in 2013 and covered separation anxiety in dogs.

Dog population management

- The Companion Animal Welfare Council (CAWC) in the UK published reports on various companion animal welfare issues, including those related to rescue, rehoming, sanctuaries and shelters.[115,116] As CAWC has now disbanded, veterinary professional associations could consider hosting these reports to preserve them, as they remain relevant today.
- Useful resources for veterinary professionals working with dogs in shelters and trap-neuter-release programmes include the Humane Dog Population Guidelines and other resources produced by the International Companion Animal Management coalition (ICAM) and the guidelines for canine catch, neuter and return programmes from JMICAW.[104]
- WOAH has created and continues to update its welfare guidelines for dog population management.[9]

Cats

11

Abstract

The core welfare concepts discussed in Part 1 are applied to cats. Common welfare hazards are listed. Common welfare outcomes are identified and structured according to the Five Freedoms framework. Practical welfare assessment protocols are referenced. Examples are provided of how veterinary professionals have advocated for cat welfare or could be more involved.

Introduction

Domestic cats (*Felis sylvestris catus*) are found worldwide in various contexts. They have a long history of domestication, with evidence suggesting that they were kept as pets in ancient Egypt, and have been associated with humans for approximately 10,000 years. They are descended from the African wild cat (*Felis sylvestris lybica*). Unlike dogs, they have not been selectively bred by humans for specific roles. Instead, they are likely to have co-evolved, being tolerated or welcomed by people for their role in controlling rodents. Breeding cats for their appearance is a relatively recent venture, starting around 150 years ago. Current attitudes toward cats vary within and between societies. For instance, they may be treasured as companions, appreciated for their role in controlling rodents, disliked as destroyers of wildlife or viewed as nuisances.

Throughout this chapter, cats will be broadly classified as either 'household cats' or 'free-roaming cats'.

Household cats

Household cats are companion animals – they live with and are cared for by humans, typically spending some or all of their time in a human home, and are 'owned' by specific people. Household cats may be kept alone or with one or more other cats. They may be perceived as lower-effort companion animals than dogs; for example, they do not require regular walks and can be kept exclusively indoors using litter trays.

DOI: 10.1079/9781786391582.0011

Free-roaming cats

The term 'free-roaming cats' encompasses strays, community and feral cats. Stray cats are those that were previously cared for as companion animals, including those that have become lost or been abandoned by their owners. Community cats are those living freely among and tolerated by human communities but not owned by any particular household. Both stray and community cats receive some food and harbourage from humans. In contrast, feral cats have little or no human contact. Categories may overlap, and cats may move between them according to changing circumstances.

Global estimates are around 350 million household cats and 480 million free-roaming cats.[117] To date, there has been little scientific study of cat welfare and large gaps in evidence exist. Most studies have focused on cats in shelters, rehoming centres and laboratories rather than cats in human homes or free-roaming cats. It appears that cats can adapt to living alongside humans in various contexts, retaining a worthwhile quality of life, provided there are some safeguards that mitigate the common welfare risks.

Common welfare hazards

- Feeding regime.
- Breed standards.
- Breeding management.
- Population control approaches.
- Level of environmental enrichment.
- Interactions with other cats.
- Level of veterinary care.
- Owner knowledge of cat behaviour.

Welfare hazards are the resources (inputs) that can lead to poor welfare outcomes. The likelihood of poor welfare outcomes and their severity (i.e. the risks) can vary. Severity encompasses both duration and intensity. Common welfare outcomes are identified below, structured by the Five Freedoms framework.

Cat welfare is determined by how well cats can adapt, both physically and mentally, to their environment (see p. 8). Relevant descriptions of 'free-living' cats are provided throughout to highlight the adjustments they must make while in human care. Here, 'free-living' is defined as cats living freely in an environment with the resources for which they have evolved and making their own choices. However, such comparisons provide only part of the welfare picture; welfare science (see Chapter 2) is essential for a proper understanding of the welfare impact of different husbandry systems.

Key determinants of welfare are the knowledge, skill and motivations of people caring for the cats, as well as whether they have sufficient resources (see pp. 24–25). Where these factors are optimal, welfare will be as good as it can be within a given husbandry system. However, there are often persistent welfare issues that are challenging to resolve without altering the husbandry system itself.

Common welfare outcomes

F1. Freedom from hunger and thirst

Cats are obligate carnivores. They must consume a meat-based diet or be given a taurine supplement to avoid becoming deficient in this essential amino acid. Cats deficient in taurine suffer severe health problems.

Household cats

Household cats are largely dependent on their human caregivers. In many cultures, they are fed the meat-based diet they require. Cats with the freedom to roam outside the house may also supplement the food they are offered by hunting.

Overfeeding tends to be an issue in many household cats.[66] Highly palatable, readily digested, commercial cat food is often provided at regular intervals daily, sometimes continuously, and household cats are often more sedentary than free-roaming cats. These factors predispose to obesity, which in turn causes or exacerbates other health problems (see F3). In addition, obesity can interfere with behavioural expression (see F5), preventing cats from engaging in activities they would normally enjoy, such as climbing and jumping. The physical format of many commercial cat foods is very different from that of natural prey. Some commercial foods do not provide enough dental exercise and teeth cleaning, increasing the risk of gingivitis and other dental problems (see F3). These conditions can make eating painful and, if severe, lead to hunger.

Figure 11-1. Farmyard cats with considerable behavioural freedom, whilst also benefiting from human care.

Cats need access to a continuous supply of fresh water to avoid thirst, even when fed moist foods that provide much of their daily water requirements.

Free-roaming cats

The extent of hunger and thirst in free-roaming community cats (Figure 11-2) often depends on the amount, quality and frequency of food and water provided by humans. This may be intentional or unintentional in the form of poorly secured food waste. It also depends on their hunting options and the number of cats competing for these resources. Where no food is provided by humans, cats often catch multiple small prey items daily, including birds, rabbits and rodents. Therefore, their diet can vary from good to poor.

Once captured and confined in rehoming shelters or sanctuaries, their choices become more limited, and they are dependent on what the people caring for them provide. Standards can vary greatly. If kept very confined and/or in close proximity to other cats, then they may refuse to eat due to feelings of fear and anxiety, resulting in hunger and weight loss.

F2. Freedom from discomfort

Household cats

Household cats are often allocated warm, dry resting areas within the home by their owners. They also usually have the freedom to choose sites they prefer from multiple other areas within and outside the house. However, cats may find certain scents and sounds that humans are oblivious to uncomfortable, for example, odours from certain cleaning agents or high-frequency sounds.

Figure 11-2. A free-roaming cat with little direct human support, scavenging for food.

Free-roaming cats

The comfort of free-roaming cats depends on the availability of various factors, such as shelter, shade and dry resting areas, and whether these are sufficient to protect them from climatic extremes. There needs to be sufficient resources for all the cats.

If free-roaming cats are captured and confined in rehoming centres or sanctuaries, they lose their ability to choose, being restricted to using what is provided for them, which will vary in quality.

F3. Freedom from pain, injury and disease

Pain, disease and injury arise for various reasons, and are likely to be accompanied by fear and distress. Diseases may be non-infectious or infectious, with the former largely due to inappropriate husbandry and predisposing to the latter. Traumatic events such as road traffic accidents are common causes of injury in cats that roam freely outdoors. There are now many excellent clinical texts providing details of common diseases and injuries, many of which the veterinary profession can treat to a high standard.

Level of veterinary care

A lack of veterinary treatment can lower welfare. Both household and free-roaming cats may receive too little veterinary intervention. This includes preventative interventions such as vaccination against common infectious viral diseases, control of internal and external parasites, dental care and neutering/desexing. It also includes appropriate treatments when sick or injured, and euthanasia in situations where veterinary interventions cannot provide a decent quality of life. For household cats, 'under-treatment' might be due to owners' lack of awareness, interest or ability to afford treatments. Difficulties in catching and handling their cats may leave them reluctant to visit veterinary clinics. Additionally, early signs of ill health may go unnoticed. Cats display less obvious behavioural signs of pain, disease or stress than dogs, possibly because they have evolved from a mainly solitary hunting species where alerting other cats to their pain is of limited help. However, with experience and training, veterinary professionals and owners can learn to recognise signs. Generally, no one is directly responsible and accountable for the general well-being of free-roaming cats, meaning they often receive no veterinary attention.

Conversely, 'over-treatment' can also lower welfare. In wealthier countries, household cats are at risk of excessive veterinary intervention due to the sheer range of possible veterinary investigations and treatments now available, similar to the situation in owned companion dogs. Veterinary professionals can strive to ensure that the likely longer-term benefits of any complex intervention for the individual cat outweigh any pain and distress caused by veterinary interventions.

Outdoor access

The freer the cat is to roam, the more normal behaviours it can express. However, there are welfare trade-offs: the greater the freedom, the greater the risks of accidents and injuries from road traffic accidents and fighting with other cats, as well as the increased risk of internal parasites through consuming infected prey animals. However, cats living solely within the home may be at greater risk of other harms, such as developing urinary tract disease or becoming obese.

Geriatric cats

Household cats are living longer, and there is a growing need for information on managing geriatric animals and assessing their quality of life, for instance animals that have chronic pain due to osteoarthritis or are experiencing general malaise due to kidney failure. However, there has been little research and development of welfare (quality-of-life) assessment tools specifically for cats.

Overpopulation

Overpopulation of cats, that is too many cats for the available resources and human community to care for or tolerate, often leads to pain, injury and disease. Cats are prolific reproducers. They can breed year-round and have a gestation period of around 9 weeks, producing several litters per year. Without human intervention, uncontrolled breeding leads to overpopulation. Among free-roaming cats, there will be considerable competition for basic resources such as food, water and shelter, meaning some cats cannot access these resources, and there will be more fighting. Infectious diseases can spread more rapidly through cats living at high densities, particularly if their immunity is reduced due to suboptimal feeding and living conditions. Mortality rates in kittens can be very high, and many are likely to suffer before they die.

Killing methods

Where society views cats negatively, there are additional welfare risks. For instance, cat overpopulation can create nuisance and public health risks for human communities, shaping attitudes. Inhumane killing methods such as drowning and poisoning are common in some countries, often coordinated by government authorities. If killing some cats is necessary, then veterinary professionals can provide education on humane methods. Killing cats alone is unlikely to effectively reduce free-roaming populations in the longer term, as other cats will move into the area to fill the void where resources that cats rely on are still available. In fact, it may even be counterproductive, as the cats that have avoided being killed may be less temperamentally suited to humans, creating a less friendly population that is more difficult to manage.

Desexing

In countries where the culture is for cats to be owned by particular households, uncontrolled breeding results in too many kittens for the number of available homes. Preventing cats from reproducing is key to protecting their welfare. Desexing, known as neutering (castration in males and ovariohysterectomy/'spaying' in females), is a common surgical mutilation of female and male cats. Desexing procedures inevitably inflict some pain and suffering. However, this is often mitigated by the routine usage of effective analgesia and anaesthesia, in contrast to many mutilations performed on farm animals. On balance, the welfare harms of the procedure often offset the welfare risks associated with leaving cats entire. Neutered cats cannot mate and reproduce, meaning fewer surplus kittens that communities struggle to care for. Neutered males have less tendency to roam and fight, and therefore lower risks of injury, pain and disease transmission. It is usually much harder for owners to control the reproductive behaviours of their cats than of pet dogs, given that it is the norm for many cats to roam outside freely and independently of their owners.

'Trap-Neuter-Release' programmes

Some animal charities operate programmes that catch/trap, neuter and release free-roaming, community cats, often referred to as TNR programmes. The general points relevant to welfare, set out for dogs in Chapter 10 (p.163), also apply to community cats. For instance, all persons must be competent in catching and handling, surgery must be performed to a high standard, and fitness for release must be properly assessed. Otherwise, such programmes can inadvertently harm animals. Early-age neutering may be warranted. The success of TNR programmes in stabilising populations and protecting cat welfare also depends on their acceptability to local communities.

'Kitten farms'

Cats may be bred to sell kittens as a business or hobby. Standards vary greatly from good to poor. A particular welfare risk is the large-scale, commercial breeding units ('kitten farms') working to low standards. For instance, many offer impoverished environments, with little socialisation of kittens during their critical period and insufficient preventative healthcare. A high incidence of respiratory disease is common in some of these establishments.

Breed standards

The business and hobby of breeding pedigree animals for their appearance is much more recent in cats than in dogs. Breed standards can cause welfare problems, such as heightened anxiety and fearfulness in Burmese cats and respiratory and ocular problems in brachycephalic cat breeds. Many of these cat breeds would not thrive without veterinary interventions. Campaigns for change are needed to avoid perpetuating breed-related problems.

Mutilations

Other mutilations, beyond desexing, are routinely performed in many cats, and all cause a degree of pain. Whether they can be ethically justified depends partly on the balance of welfare benefits and harms for the cat. For instance, microchipping is a minor, momentary procedure that provides permanent identification and has the major benefits of enabling lost cats to be reunited with their owners.[120] It links cats with specific people, keeping those owners accountable for the animals' well-being. In contrast, the mutilation of 'declawing' is much more invasive and painful. Here, the cat's claws are surgically removed for the owner's convenience to prevent normal scratching behaviour that can damage household furniture. This procedure remains legal in many countries, including the USA, and is difficult to justify ethically. (Declawing for non-medical reasons is illegal in the UK.)

F4. Freedom from fear and distress

Unpleasant feelings, such as fear, often occur alongside pain, injury and disease. Fear and distress also occur independently. To protect the cat's welfare, these feelings need to be kept short-lived and at a level the cat can readily cope with. Examples of situations that cause fear and distress are as follows.

Social relations with other cats

Cats may become fearful and distressed when forced to live near other cats they prefer to avoid. 'Free-living' domestic cats tend to live alone or in small groups, called colonies, with group members typically consisting of several related females and their kittens. They do not readily accept new adults into the group, and males live separately on the periphery of established groups. In contrast, human owners decide the social grouping of household cats. Even when kept singly, cats may feel stressed by a high density of household cats in their neighbourhood. Contact does not need to be face-to-face to affect the cat. They communicate using a combination of visual, tactile, auditory and olfactory signals, with their excellent sense of smell being particularly important for communicating with each other when they are apart. The scent of another cat brought into the house unknowingly by a person, for instance on a piece of furniture, is enough to signal the presence of another cat.

Chronically stressed cats often exhibit only subtle behavioural changes that are not immediately apparent to owners. Examples are eating less and reduced activity levels. Sometimes, they show abnormal behaviour, such as overgrooming. They may also show overt behavioural responses, such as toileting in the house when they are unable to access litter trays dominated by other cats. They will also spray urine inside the house as a scent mark to reassert their territory. Owners may not understand cat behaviour, and therefore not address the underlying causes. They may compound the stress by punishing the cat. In addition, cats experiencing fear and anxiety are more prone to physical conditions such as urinary tract disease.

It is essential to understand that cats are selectively social. In multi-cat households, providing sufficient resources – including food and water stations, litter boxes and enrichment options – prevents one cat from dominating. Creating a safe environment with hiding places and physical barriers allows cats to retreat when needed. Cats also enjoy climbing to elevated spots for observation. Lastly, maintaining a consistent daily routine and ensuring positive interactions with humans and other pets can help cats feel more secure.

Note that cats tend not to show 'separation anxiety', being a more solitary species than dogs.

Socialisation

Kittens have a 'critical' or 'sensitive' period between 2 and 7 weeks of age during which they need to be socialised, that is exposed to different people, animals and environments in a way that develops positive associations. Properly socialised cats can learn to accept other cats and species, such as dogs and humans, and to find their company enriching. In contrast, a lack of proper socialisation leads to lifelong fears and anxieties. Cats from impoverished environments, such as those on 'kitten farms', are at particular risk. Unresolved 'behavioural problems' can lead to cats being relinquished to rehoming centres. They already have psychological issues and are then further burdened with the fear and anxiety caused by confinement.

Relinquishment to rehoming centres

Confining large numbers of cats in close proximity in rehoming centres and shelters is likely to cause many of them some stress and anxiety. The extent to which the cat copes will depend on the standards within the individual shelter, for instance the skill of the carers and whether enrichments that allow expression of important normal behaviours are provided.

Any rehoming strategies need to be suitably nuanced. Protecting welfare involves understanding the individual cat's personality and life experiences. Stray cats with prior experience living in a household may be suitable for rehoming. However, cats that have always been free-roaming are unlikely to adjust to life as a household cat, and attempting to do so will replace one form of welfare harm with another. Releasing neutered individuals back into their familiar environment, but with an added level of ongoing support, is often the best option in communities where this is considered acceptable. For example, some cats could be placed as free-roaming farm cats for rodent control, with human guardians allocated to monitor them and ensure their long-term care and welfare (Figure 11-1).

Veterinary clinic visits

Visits to the veterinary clinic can be particularly fear-inducing. However, there are ways to make the veterinary practice environment more 'cat-friendly', reducing feelings of threat. Even small changes can be helpful, such as providing a hospitalised cat with a hiding place, such as a cardboard box, and a familiar item carrying the scent of their home, such as a commonly used blanket.

F5. Freedom to express normal behaviour

Ethograms for domestic cats show they retain many of the normal behaviours of their ancestors, the African wild cat. While individuals have different personalities, experiences and preferences, some behaviours are common to all cats.

Some normal behaviours important to the cat have already been discussed in relation to the other Freedoms. Additional normal behaviours cats are motivated to engage in include hunting, climbing, jumping, scratching objects, exploring, resting at vantage points, retreating to shelter and hiding. They also rub on objects to deposit scent and spray urine. Many of these behaviours are likely to enhance the cat's quality of life, providing enjoyment and keeping them fit. Others increase their feelings of security, or are important for their communication with other cats, including establishing territories.

Hunting

Domestic cats remain motivated to hunt when the opportunity arises, even if they are not hungry or other food is freely available. They are attracted to sudden movements of small creatures and dark crevices where rodents are likely to hide. Cats have very sensitive hearing, and their mobile funnel-shaped pinnae help them locate sound sources accurately. They are thought to have the widest hearing range of any domesticated mammal and can detect ultrasound emitted by rodents.

Communication

In particular, scent marking enables cats to communicate without ever physically meeting. Cats signal to each other by depositing scent marks when they urinate, defecate, spray urine or rub on objects or each other. Spray marking is a distinct behaviour where cats back up to a vertical surface and eject urine backwards whilst quivering their tail. Both sexes perform this behaviour, though it is more common in males. It is important for marking territories. Cats also deposit scents from their interdigital glands when they scratch objects, regularly scratching vertical surfaces, such as tree trunks and wooden posts. Scratching behaviours also condition their claws.

Behavioural autonomy (choice)

The degree to which different cats can engage with these behaviours depends largely on the choice their environment affords. 'Free-living' cats often have considerable behavioural freedom and potential for a good quality of life, provided sufficient resources, such as space, food and shelter, are available for all the cats. At the other end of the spectrum, people tightly control the lives of cats kept confined indoors all of the time, for instance, determining where they rest, exercise and when they eat. The extent of normal behavioural expression depends on the enrichments provided by the owner, which in turn determines whether they become bored and frustrated or are well adjusted.

Some cats adjust to indoor living, but it does not suit all cats, and their quality of life needs to be weighed against any risks they face.[121] Also, cats may differ over the course of their lifetime. For example, elderly or disabled cats may prefer more time indoors.

Environmental enrichment

Changes are only enriching if they are meaningful to the cat. For example, climbing frames that provide access to high shelves, platforms or window sills are appreciated, as are comfortable, soft areas for resting at varying heights. Materials cats can use for scratching, such as wooden or sisal scratch posts, rush matting, or pieces of carpet, are important. Humans can also enrich their cat's life by interacting with them in a way that matches their temperament. For example, some individuals prefer being petted and groomed, while others enjoy interacting through play with toys. Any toys should be of value to the cat, and not just look good to humans. Examples include toys that move and mimic prey characteristics to recreate hunting behaviours. Cats also like to explore, so novel objects like boxes can be introduced intermittently. These also provide hiding places and a sense of security. Although cats enjoy some novelty and exploring, they also like their daily routine, environment and interactions with humans and other animals to be somewhat predictable, and to avoid sudden, major changes.

As already mentioned, cats have a more acute sense of smell than people, and different detergents that seem innocuous to us can transform a cat's world (see F4). For instance, cats may avoid using litter trays that have been cleaned with disinfectants they find aversive. It would be helpful to know much more about how other sounds and scents in the home affect a cat's well-being.

Welfare assessment protocols

Validated tools for assessing cat welfare are limited, and little research has been conducted in this area.[122] Existing welfare assessment tools primarily focus on health status. Examples include the Karnofsky score modified for cats,[123] the CatQoL tool proposed by Bijsmans, and the Cat Health and Wellbeing (CHEW)

Questionnaire.[124] The modified Karnofsky score for cats, which is adapted from human medicine, consists of two parts: Part 1 is a clinician's evaluation of the cat's physical health, and Part 2 is an owner questionnaire that includes behaviour-based questions regarding the cat's quality of life. The CatQoL tool is a questionnaire designed to compare the quality of life of cats with chronic kidney disease with that of healthy cats. The CHEW Questionnaire is intended for completion by cat owners to measure their cat's general quality of life. All of these assessment tools require further research to validate them.

Validated, practical assessment tools focusing on measuring acute or chronic pain are already available, highlighted in the WSAVA Global Pain Guidelines.[44]

Most of the current tools are intended for use by cat owners in home environments. However, there is currently no comprehensive welfare evaluation tool available for cats housed in rehoming centres or sanctuaries.

Advocacy examples

- The International Society of Feline Medicine (ISFM), which is the veterinary division of the charity International Cat Care (ICC) has published a 3-part series titled 'Feline Behaviour and Problem Behaviours' in their society journal, which can be accessed free of charge.[125] Other ISFM veterinary resources are freely available, including posters, infographics and videos on important welfare topics, such as feeding, enrichment, scent communication among cats, as well as equipment needs and layouts for households with multiple cats.
- The ICC has collaborated with the BVA on the HealthOverLooks campaign, which urged cat lovers to avoid choosing pedigree breeds with extreme anatomical features.
- The ISFM provides a 'cat-friendly clinic' website to help reduce stress for both cats and their owners during visits to veterinary clinics. For example, this resource emphasises gentle restraint methods and discourages forceful techniques, such as scruffing. It also includes educational videos on cat handling for veterinary staff and owners.
- Valuable sources of information include accessible texts by credible behaviourists, such as the book 'Cat Sense' by John Bradshaw,[126] which can benefit owners and veterinary professionals alike.
- The creation of official guidance, such as the UK Code of Practice for Cats, published in 2017 under the Animal Welfare Act 2006, is important for educating owners about their legal responsibilities and how to meet these.
- There are many useful resources available from the International Companion Animal Management Coalition (ICAM),[58] such as guidelines on humane cat population management and the development of euthanasia policies.

- In the UK, veterinary professionals have contributed to and can consult the PDSA Pet Animal Wellbeing (PAW) reports[66] and the various Companion Animal Welfare Council (CAWC) reports, which helped establish priorities for further advocacy.
- Veterinary professionals can support initiatives such as the Animal Behaviour and Training Council's efforts in the UK to strengthen the accreditation process for its members.
- Microchipping cats is important for their welfare, as it helps reunite lost, stray or escaped cats with their owners, and is endorsed by ICC and its veterinary division. In some countries, such as the UK, microchipping of cats has become a legal requirement following strong campaigning.

Rabbits

12

Abstract

The core welfare concepts discussed in Part 1 are applied to rabbits. Common welfare hazards are listed. Common welfare outcomes are identified and structured according to the Five Freedoms framework. Practical welfare assessment protocols are referenced. Examples are provided of how veterinary professionals have advocated for rabbit welfare or could be more involved.

Introduction

In the UK, many veterinary professionals encounter rabbits (*Oryctolagus cuniculus*) as companion animals, and some encounter them as laboratory animals. Many are probably less aware of the large numbers of rabbits farmed primarily for their meat in other countries, including Europe, with Spain, France and Italy being particularly large producers. Human actions also impact the welfare of wild rabbits.

In the UK, rabbits are the third most popular mammalian companion animal after dogs and cats. Around one-quarter of owners reported acquiring their rabbit primarily as a child's pet.[66] However, contrary to public perceptions, they are not an 'easy option'. Rabbits have complex social structures and other needs that can be difficult to provide for in captivity, and veterinary professionals in the UK have identified pet rabbit welfare as a key concern.[66]

Interest in the welfare of farmed rabbits appears relatively recent but growing. By far, the commonest commercial farming systems are highly intensive, with rabbits housed in barren cages. Rabbits are highly productive due to their short gestation and lactation periods and great prolificacy.

Rabbit welfare research is limited but growing. To date, much of it has concentrated on laboratory rabbits, with less focus on companion rabbits and very little on farmed rabbits.

DOI: 10.1079/9781786391582.0012

Husbandry

Companion (pet) rabbits

The lifespan of well-cared-for domestic pet rabbits can be 8–12 years, depending on the breed. Traditional housing for pet rabbits consists of a hutch, usually a wooden box with a mesh window on the front, that is often located in the garden (Figure 12-1). There may be an exercise run temporarily or permanently attached. They may be housed alone or with a companion. In the UK, around 60% of companion rabbits live primarily outside the owner's home. Around 40% live mostly inside the owner's home, where they may be free-roaming or live in a cage or hutch.[66]

Farmed rabbits

Adult female rabbits (does) are first bred when they reach sexual maturity at around 5–6 months of age. They usually undergo artificial insemination after induced ovulation. The gestation period lasts about 30 days, after which the doe gives birth to a litter of kits. The kits are weaned at 35 days old. The doe is inseminated a few hours or days after weaning. She produces a new litter approximately every 6 weeks over a productive lifespan of typically one year, producing up to 30 kits in this time, after which she is culled.

Does are housed with their kits from birth until weaning, and are kept alone at other times (Figure 12-2). After weaning, the offspring are referred to as 'fattening rabbits' or 'fatteners'. They are slaughtered when they reach a weight of around 2.5 kg at 70–80 days of age.[127]

Figure 12-1. A typical style of rabbit hutch used for pet rabbits in the UK

Figure 12-2. A typical production system for rabbits farmed for meat. Highly intensive cage systems are the norm, with small, barren wire cages. Urine and droppings fall through the wire cage floor.

In European farming systems, 'fatteners' are usually housed as groups of 6–8 animals in small cages, approximately 30 cm in height.

Breeding does are usually housed individually in small wire cages of approximately 60 cm in height, except after parturition, when they remain with their litter until weaning. At parturition, nest boxes are often provided. Sometimes cages contain a platform so the doe can withdraw from her litter, once they start to leave the nest box.[127]

Some alternative systems with better welfare potential are being developed, but are still niche.[128] Included are 'enriched cages' that contain a platform and gnawing material, often in the form of a wooden stick; 'park' housing, where larger cages hold around 20 rabbits, with a platform and gnawing material; and pen housing where one or more rabbits are kept in an open-topped enclosure, which may be indoors or outdoors.

Common welfare hazards

- Feeding regime.
- Level of confinement and housing design.
- Space allowance.
- Level of environmental enrichment.
- Social groupings.

- Thermal environment and air quality.
- Level of veterinary care.
- Handling quality.

Welfare hazards are the resources (inputs) that can lead to poor welfare outcomes. The likelihood of poor welfare outcomes and their severity (i.e. the risks) can vary. Severity encompasses both duration and intensity. Common welfare outcomes are identified below, structured by the Five Freedoms framework.

Rabbit welfare is determined by how well rabbits can adapt, both physically and mentally, to their environment (see p. 8). Relevant descriptions of 'free-living' rabbits are provided throughout to highlight the adjustments they must make while in human care. Here, 'free-living' is defined as rabbits living freely in an environment with the resources for which they have evolved and making their own choices. However, such comparisons provide only part of the welfare picture; welfare science (see Chapter 2) is essential for a proper understanding of the welfare impact of different husbandry systems.

Key determinants of welfare are the knowledge, skill and motivations of people caring for the rabbits, as well as whether they have sufficient resources (see pp. 24–25). Where these factors are optimal, welfare will be as good as it can be within a given husbandry system. However, there are often persistent welfare issues that are challenging to resolve without altering the husbandry system itself.

Common welfare outcomes

F1. Freedom from hunger and thirst

Rabbits are herbivores, evolved to eat and digest low-energy, high-fibre foods. They spend around 70% of their waking time foraging and eating. Their teeth grow continually throughout their lives and are constantly worn down by the long periods spent chewing vegetation. They extract maximal nutrition from fibrous food by performing caecotrophy. Here, food passes through the gut twice; the first time, it produces special droppings, called caecotrophs, that are reingested.

The diet of companion and farmed rabbits is frequently inappropriate. For instance, around 20% of pet rabbits in the UK are fed rabbit 'muesli', a mix of seeds and flakes,[66] as part of their main diet, predisposing them to obesity as they select the high-sugar components over fibrous elements. Around 85% of owners feed treats, some of which have a high sugar content. An estimated one-quarter of pet rabbits are overweight, and many owners are unaware of their rabbits' weight or body condition score. In the UK, only around three-quarters of rabbits are fed any long fibre as part of their main diet, and one-third are offered too little.[66] Ideally, at least 80% of their daily food intake should be good quality, long fibre, and rabbits should have access to this day and night. Examples include dust-free hay or fresh grass, but not grass clippings. Lack of fibre leads to inadequate tooth wear, which commonly progresses into dental pathology, where the rabbit finds eating painful and eats less, becoming thin and malnourished.

Farmed rabbits are routinely fed solely on an ad-lib pelleted diet, completely lacking any long fibre. While this diet provides the macro- and micronutrients required for production, it fails to meet their welfare needs, leading to various dental, digestive, and respiratory health problems, as well as unpleasant emotions such as boredom and frustration, by preventing important normal behaviours.

F2. Freedom from discomfort

The environments of captive rabbits are much less complex than those of 'free-living' rabbits and, in many cases, limit their basic comfort behaviours. Rabbits prefer to live in dry areas, with soft and sandy soil ideal for burrowing. They dig large, complex burrow systems with numerous interconnecting tunnels, known as warrens, where they live as stable, social groups of between two and 20 individuals. Each warren can accommodate several social groups, totalling as many as 50 individual rabbits or more. They generally have considerable space and choice of resources within their immediate environment, sufficient for all group members. For example, they can find comfortable, dry resting areas and keep their toileting areas entirely separate from these. However, many captive rabbits do not have these choices. Around one-quarter of pet rabbit owners in the UK chose hutches smaller than the minimum recommended sizes.[66] Whilst some pet rabbits may be offered bedding and additional space as an outdoor 'run', most farmed rabbits are continuously confined in small, barren cages that severely limit their movement. They rest on wire floors lacking bedding and often develop skin sores as a result.

F3. Freedom from pain, injury and disease

Injury and disease have multifactorial causes and are often accompanied by pain. Rabbits show subtle pain signs, such as a slight change in posture, facial expression and ear position, which can easily be overlooked.

Feeding regime

Inappropriate diets, particularly those lacking long fibre, are common in captive rabbits. They prevent different body systems from functioning properly, manifesting as dental, digestive and respiratory disorders. Dental problems are very common due to inadequate wear of the continually erupting teeth.

The rabbit then has difficulty eating and, as the problem progresses, develops mouth lacerations and abscesses, which cause considerable chronic pain.

Confinement

Many captive rabbits are confined to small spaces, limiting their ability to exercise. This lack of weight-bearing activity can lead to painful musculoskeletal conditions, such as osteoporosis and related fractures. Companion rabbits kept inside the house are probably less likely to be forgotten and often have more freedom to roam than those confined to hutches outdoors. However, the home environment must address various welfare risks, such as access to electrical wires that the rabbit may chew, and a lack of quiet, secure, non-frightening areas.

Fighting

Choosing social companions for captive rabbits can be challenging, in particular avoiding fighting and associated injury. 'Free-living' rabbits are highly social, living in stable groups comprising males and females, often from the same family. Group members form close social bonds and establish a clear dominance hierarchy. However, at times, this hierarchy is challenged, for instance as the physiological state of rabbits alters during the breeding seasons, causing them to be more territorial and aggressive. In their complex natural environment, rabbits can adapt their behaviours to escape such threats. However, captive rabbits are much more confined and have few choices.

Fighting is common when unfamiliar rabbits from different family groups are mixed, and this should generally be avoided. However, such mixing may be a routine part of the farming system. Also, in farmed rabbits, there is the additional issue of managing harmful behaviours as the reproductive state changes. Farmed rabbits must be kept entire (not desexed) because the females are required to breed, and neutering the fatteners is impractical. Fatteners are usually housed as small groups of familiar individuals, which, before sexual maturity at around 12 weeks of age, can be peaceful and allow for some positive social interactions. However, at sexual maturity, or even earlier regarding rank setting, relationships change, and it is common for them to inflict serious fight wounds on each other.[127] To avoid fighting, breeding does tend to be housed in isolation, which leads to different welfare harms.

Mutilations

Companion rabbits are generally not required to reproduce. Surgical neutering of males and females greatly limits territorial and aggressive tendencies in both same-sex and mixed-sex groups. In addition, neutering means that males and females can be kept together whilst avoiding unwanted kits. Neutering in this context has overall welfare benefits, but it is still an invasive, painful and fear-inducing surgical procedure. Mitigations include good practice in handling, anaesthesia and analgesia.

Level of veterinary care

The welfare of captive rabbits may be compromised by insufficient veterinary intervention. For instance, there are limited authorised veterinary medicines, both preventatives and treatments, for rabbits as they are a minority species and not financially worthwhile for pharmaceutical companies to invest in.

Companion rabbit owners may be unaware of or unable to afford veterinary care despite the recent growth of medical and surgical options. For instance, despite readily available vaccines, under half of UK pet rabbits were found to have been vaccinated against serious viral diseases that cause suffering and death, such as myxomatosis and viral haemorrhagic disease.[66] Farmed rabbits may be vaccinated for economic reasons, but will receive limited or no individual veterinary attention when they become injured or ill.

Morbidity and mortality

Morbidity and mortality levels in farmed rabbits tend to be higher than in other farmed species, reported from 10–15% between birth and weaning and 5–15% between weaning and slaughter.[127] Mortality is highest around parturition and during the first week after weaning, both times of high stress that make the animals more susceptible to infections. Mitigations include strict preventative hygiene and optimising housing, ventilation and management. However, economic limitations dictate high stocking densities and severe behavioural restrictions, so high stress levels remain the norm.

Breed standards

Some breeders of companion rabbits have prioritised appearance over health and welfare. Examples include the extreme conformation of the Netherland dwarf and Lionhead breeds, selected for brachycephalism and a longer bottom jaw, which causes dental malocclusion and leads to dental pathology. Brachycephalic rabbits also typically have 'lop' rather than erect ears, predisposing them to middle ear infections and interfering with their hearing, disrupting normal communication.

F4. Freedom from fear and distress

Fear and distress are often associated with pain, injury and disease, but they can also occur independently. Momentary fear is a normal part of life and a protective mechanism. However, prolonged or intense fear from which the rabbit cannot escape harms their welfare. Common factors causing fear and distress in captive rabbits include social isolation and exposure to real or perceived predators.

Social isolation

Rabbits are a social prey species, highly motivated to maintain social contact with other rabbits, with studies showing this is as important to them as accessing food when hungry.[129] Social isolation is highly likely to induce fear and anxiety. Conversely, studies show that rabbits housed with familiar, friendly individuals spend more time engaging in behaviours indicating they are relaxed, such as grazing, resting and playing, rather than being alert for threats. However, many companion rabbits and farmed breeding females are kept alone.

Social stress

Providing companions for companion rabbits is often practical, though they must be carefully chosen and introduced to make sure relations are positive and to avoid fighting. Suitable companions include neutered rabbits of a similar size or neutered litter mates (guinea pigs have traditionally been chosen but are not recommended because they have different dietary requirements, can be injured by rabbits, and each requires the company of its own species). However, providing companions for breeding females on conventional farms is much more challenging. When group-housed, they tend to fight, severely lacerating each other's skin, and they disturb each other's nests, causing high kit mortality. Therefore, farmed breeding females are caged alone, with the only company being their kits between parturition and weaning. In contrast, 'free-living' does spend much of the day with other group members, only separating themselves to give birth and feed their kits. Therefore, some alternative, niche systems aim for higher welfare by using a 'park' system.[127] Here, does are only individually housed at parturition, being housed with other does after their litter is weaned. However, the welfare of rabbits in these systems needs to be studied scientifically to establish whether the purported benefits are being delivered.

Perceived threats

Captive rabbits, whether companion or farmed, are often exposed to various sights, sounds, smells and physical touch that they perceive as threatening, and their environment often lacks the space and resources they need to take effective action to alleviate their fear.

Rabbits live in a very different sensory world from humans, who may be oblivious to the information their rabbit is receiving and its impact on their well-being. Rabbit senses have evolved to detect predators in their natural environment early and accurately. They have excellent hearing, and their large mobile funnel-shaped pinnae help them pinpoint sound sources. Their vision has evolved to scan the environment, being panoramic and highly sensitive to movement. They rely greatly on smell for detecting predators and obtaining other information about their environment. Scent marking is a major means of communication, and they mark objects by chin rubbing, depositing faeces and spraying urine.

Other common companion animals, such as dogs and cats, are natural predators for rabbits, and their proximity often creates fear. Rabbits are also predisposed to view people as potential predators, and so the quality of human–animal interactions is very important for counteracting this. Rabbits have good learning abilities and can habituate to human contact and even other pet species, finding interactions rewarding. Proper socialisation at a young age is key. However, certain types of handling will always remain innately frightening to them, and owners can easily misinterpret certain fear behaviours. For instance, rabbits will find being lifted to a height very fear-inducing, mimicking being lifted by a predator. Rabbits that 'freeze' lying motionless when placed on their backs are very frightened, not calm or 'hypnotised'.[130] Unlike dogs and cats, rabbits tend not to vocalise unless very distressed. Many owners do not realise that rabbits thump their back feet to signal danger and that this behaviour means they are probably fearful or anxious about something in their close environment or find handling aversive.

The environment of captive rabbits often lacks the space and resources needed for taking effective action to alleviate their fear. 'Free-living' rabbits limit their exposure to predation by spending much of their daylight time in underground burrows and being active above ground mainly during dawn and dusk. Where captivity confines them to a barren space without places to withdraw and hide, they are likely to be left feeling exposed and vulnerable, creating chronic fear. Similarly, they cannot perform important vigilance behaviours such as standing tall on their hind legs to enhance the visual scanning of the environment.

Educating owners has the potential to improve the welfare of many pet rabbits. However, specific logistical challenges in farming are economically difficult to address. There are also serious welfare issues associated with handling, transporting and slaughtering farmed rabbits, many of which have only recently received attention.

F5. Freedom to express normal behaviour

Domestic rabbits retain many of the behaviours of their ancestors, the wild European rabbit (*Oryctolagus cuniculus*).[127] 'Free-living' rabbits spend large parts of their waking time foraging and eating, and they tend to gnaw on objects. Other behaviours they are motivated to perform daily include running, jumping, hopping, digging, resting, exploring, standing up on their hind legs, withdrawing and hiding, and interacting socially with others in their group, preferring to synchronise their behaviours with other group members. They also prefer to use different areas of the living environment for different purposes, such as feeding, resting and excretion. Young rabbits are particularly active and playful.

Environment lacking variety

The environment provided for captive rabbits is often too small and barren to allow them to express important behaviours. Rabbits denied these are likely to experience unpleasant emotions such as frustration and boredom, and be physically unfit. They may redirect their behavioural expression to other less preferred substrates; for instance, house rabbits may dig at carpets and chew furniture. Some rabbits develop 'abnormal' behaviours, including stereotypic behaviours, such as repetitive bar-biting, water bottle chewing and self-mutilation. Also, the complex social group structures of 'free-living' rabbits (see F3 and F4) are often difficult to cater for in captivity in a way that maintains social harmony.

Environmental enrichment

The scope for improvement varies with context. For pet rabbits, veterinary professionals can help mitigate problems by educating owners on how to provide for fuller behavioural expression and the enjoyable company of other rabbits and humans. It is often practical and affordable to introduce enrichment items such as tubes to hide in, a large litter tray filled with potting compost to explore and items to investigate and gnaw on, such as wooden logs, as well as ad-lib access to suitable long fibre in the diet. The diet is as important for mental health as it is for physical health. Appropriate social companions have already been discussed.

To date, there has been relatively little scientific study of enrichments, and those undertaken have focused on laboratory rabbits. They indicate that those kept in enriched cage systems are less stressed than those in barren cages, because they are less restless, less timid, gnaw the bars less and tend not to groom excessively.[131] Welfare interest and study in laboratory rabbits are helping to create impetus for change in other sectors. By law, most laboratory rabbits in EU countries must be allowed to eat hay, jump on a platform, easily sit, move and lie, and interact socially with other litter mates. In contrast, before 2013, most were housed alone in barren cages.

Owner knowledge

Welfare harms are likely exacerbated by many rabbit owners not understanding normal behaviour, especially where the behaviour causes them inconvenience. In the UK, around half of pet rabbit owners reported that their rabbits displayed at least one behaviour that they would like to change, with the top three behaviours being thumping of the back feet, digging up ground and/or carpets and chewing furniture.[66] As a result, owners may reduce the frequency of their interaction with the rabbit or punish the rabbit, which is unpleasant for the animal and can worsen stress-related behaviours. 'Problem behaviours' are partly responsible for some 60,000–70,000 rabbits being relinquished to rehoming centres in the UK.[132] Some owners may request euthanasia.

Owners can also learn to use rabbits' good learning abilities so they better fit around their lifestyles. For instance, they can be taught to respond to commands and house-trained using positive rewards. When 'problem behaviours' occur, the UK Animal Behaviour and Training Council[133] maintains a register of behaviourists, helping to locate those who use humane, evidence-based methods.

Logistical and financial constraints

Farmed rabbits have similar needs to pet rabbits, but the challenge is providing improvements that work in a commercial environment. For instance, enrichment items need to be safe, easy to clean, as well as meaningful for the rabbit and sustaining their interest. They bring considerable extra work and, therefore, cost. Currently, most farmed rabbits are kept in conventional, small, barren cages where performing many normal behaviours is impossible. Some cages for breeding females are referred to as 'enriched', but whilst better than conventional cages, they are still minimal – they provide a stick to gnaw on, there may be an area of solid floor rather than just wire and sometimes a raised platform is provided. Caged fatteners have more space for locomotion than breeding does, but it is still inadequate, and they rest for around half of their time or more, which is much longer than in free-living rabbits.

Welfare assessment protocols

Rabbit protocols appear to be at an early development stage, but research in this area is increasing.

Farmed rabbits were not included in the AWIN or Welfare Quality projects (see Chapter 3). In 2022, the French Agency for Food, Environmental and Occupational Health & Safety (ANSES) released a helpful document titled 'List of Welfare Indicators and Methods of Assessment for Rabbits on Farms'[134] that suggests a range of measures that could form the basis of protocols.

Advocacy examples

- Rabbit Awareness Week in the UK, organised by the Rabbit Awareness Action Group, offers free educational resources for rabbit owners, vets, retailers, and rescue/rehoming centres. Veterinary practices often support the initiative in various ways, such as providing free rabbit consultations and discounts and disseminating information on their social media feeds. Each year features a different theme, such as the environment in 2022, neutering in 2023, and a healthy diet in 2024.
- The UK charity, the Rabbit Welfare Association and Fund[132] runs various campaigns, including a call for rabbit breeders to require licensing, not dissimilar to dog breeders.

- Growing awareness of rabbits led to the publication of an updated Good Practice Code for the Welfare of Rabbits in 2024, issued to support the Animal Welfare Act 2006.
- In the UK, the BVA, British Veterinary Zoological Society (BVZS) and the British Small Animal Veterinary Association (BSAVA) joint policy recommends that companion rabbits be housed in compatible pairs or groups and guides the veterinary profession in helping owners achieve this.[135]
- Improving the welfare of farmed rabbits poses different challenges than for companion rabbits. There needs to be greater societal awareness of how rabbits are raised for meat to create momentum for change. The FVE position statement regarding farmed rabbits is a good starting point.[127]
- Further research is needed into rabbit welfare and how to provide them with a decent quality of life in captivity. Surveys like the PDSA Animal Wellbeing (PAW) reports, conducted annually by the PDSA[66] in the UK, are very useful for benchmarking rabbit welfare and directing efforts and resources where they are most needed.
- While awaiting sufficient context-specific research, decisions can be guided by the best available evidence, primarily from studies on laboratory rabbit welfare. Useful resources include guidance on housing, husbandry and handling found on the 'NC3Rs' website.[136]
- The EU animal welfare reference point for poultry and other small farmed animals (EURCAW Poultry SFA) has been established[137] and will be providing resources and training for official inspectors who enforce rabbit welfare standards.

Horses

13

Abstract
The core welfare concepts discussed in Part 1 are applied to horses. Common welfare hazards are listed. Common welfare outcomes are identified and structured according to the Five Freedoms framework. Practical welfare assessment protocols are referenced. Examples are provided of how veterinary professionals have advocated for horse welfare or how they could become more involved.

Introduction

Horses (*Equus ferus caballus*) are kept for many different purposes, which can be broadly categorised as 'working' or 'leisure', though the boundary between them is somewhat blurred. Working horses fulfil many roles globally, such as pulling carts, carrying goods, transporting people and being used by institutions such as the police or army. They are used in sports such as professional racing, show jumping, dressage, eventing, polo and driving, as well as in horse breeding businesses. In some countries, they may be reared for their meat or even milk. Leisure horses are often companion animals, though they are frequently used in amateur versions of the above-listed sports.

Research into the welfare of horses has greatly increased in recent years, especially in Europe. The three most studied aspects appear to be stress and pain evaluation, social behaviour and human–horse relations.[138]

Horses from different sectors are of different breeds and types and managed quite differently, and there is considerable nuance within each sector, meaning that the welfare risks also vary somewhat. Generally, many husbandry practices have developed out of tradition and convenience, rather than from an understanding of horse behaviour and welfare.

This chapter illustrates core welfare concepts, mainly using examples from leisure horses in European countries. However, the welfare of horses in any sector can be analysed from first principles using a similar approach.

DOI: 10.1079/9781786391582.0013

Common welfare hazards

- Feeding regime.
- Type of activity or work.
- Level of confinement.
- Social groupings.
- Mixing unfamiliar horses.
- Level of hoof and dental care.
- Parasite management.
- Level of veterinary care.
- Suitability and maintenance of tack and equipment.
- Handling quality and training methods.

Welfare hazards are the resources (inputs) that can lead to poor welfare outcomes. The likelihood of poor welfare outcomes and their severity (i.e. the risks) can vary. Severity encompasses both duration and intensity. Common welfare outcomes are identified below, structured by the Five Freedoms framework.

Horse welfare is determined by how well horses can adapt, both physically and mentally, to their environment (see p.8). Relevant descriptions of 'free-living' horses are provided throughout to highlight the adjustments they must make while in human care. Here, 'free-living' is defined as horses living freely in an environment with the resources for which they have evolved and making their own choices. However, such comparisons provide only part of the welfare picture; welfare science (see Chapter 2) is essential for a proper understanding of the welfare impact of different husbandry systems.

Key determinants of welfare are the knowledge, skill and motivations of people caring for the horses, as well as whether they have sufficient resources (see pp. 24–25). Where these factors are optimal, welfare will be as good as it can be within a given husbandry system. However, there are often persistent welfare issues that are challenging to resolve without altering the husbandry system itself.

Common welfare outcomes

F1. Freedom from hunger and thirst

Horses require sufficient quantity and quality of feed and water to meet their nutritional needs. They are grazing animals, evolved to consume high-fibre vegetation frequently throughout the day. When given the choice, they typically spend about 60% of their daylight hours moving while grazing, covering significant distances, and consuming a variety of grass species. A core part of their digestion involves fermentation in the large intestine. When water is readily accessible, they typically drink between 20 and 70 litres per day, divided into several drinking sessions. The amount increases if they are late in pregnancy, lactating, or when ambient temperatures are high.[139]

Diets of captive horses may not meet the horse's nutritional needs. Common root causes are owners lacking nutritional knowledge and difficulty accessing feed/pasture, due to limited availability or financial constraints.

Extensive group grazing most closely resembles the environment horses have evolved for and often meets their dietary needs. However, factors such as a lack of rainfall and overgrazing can reduce the quality and quantity of grazing, leading to hunger. Also, grass growth in countries such as the UK is seasonal, stopping in the winter, and owners may fail to supplement grazing with hay or haylage at this time.

Even where pasture is of good quality, grazing alone may not meet the higher nutritional requirements of horses that are very active, growing, pregnant or lactating, and they will be undernourished without appropriate supplementary feed.

Grazing is commonly supplemented with commercial, cereal-based feeds. The diet must be carefully managed to ensure the horse still consumes sufficient fibre-based elements and retains proper digestive function. Where this goes wrong, common health issues are dental disorders, laminitis and colic (see F3). In turn, horses suffering from dental disorders can struggle to eat sufficient food.

'Free-living' horses are social animals that typically live in stable groups (herds) with a well-established social hierarchy, which maximises social harmony and allows them to resolve their differences with little overt aggression. They synchronise many of their behaviours, including feeding. To maintain friendly relations, it is essential to provide captive horses with adequate space and strategically plan feed locations to ensure all horses in a group have access. Without this, excess competition can leave subordinate horses hungry. If management practices lead to excessive competition, there can be significant social stress and physical injuries from fighting (see F3 and F4).

Obesity due to overfeeding is common in many leisure horses in wealthier countries. Owners may lack knowledge of, or training in, body condition scoring and thus normalise obesity.

F2. Freedom from discomfort

The natural habitat of horses is usually varied, allowing them to seek out comfortable resting areas. 'Free-living' horses typically spend around one quarter of their time resting, mainly while standing but also lying down. They sleep while lying down, usually during the night. They commonly seek resting spots with dry ground to lie on and where they are less exposed to insects that irritate them. Horses can generally adapt to a wide range of environmental conditions by seeking shelter and shade and altering their behaviour accordingly, for instance changing posture and activity levels to regulate heat loss. Other less immediate adaptations include changes in the amount of coat cover that insulates their body according to climate and season.

The comfort levels of captive horses vary according to the environment and resources provided by people, which determine their options for getting comfortable. Manmade structures, such as shelters in fields, can provide additional protection against inclement weather and shade in hot weather, supplementing natural protections such as trees and hedges. There needs to be sufficient space for all group members, or subordinate individuals may be displaced and uncomfortable. Confining horses indoors in stables can offer additional protection from adverse weather and insects and facilitate the provision of supplementary feed and comfortable dry bedding. However, prolonged confinement in the less interesting stable environment can lead to different welfare issues (see F5 and Figure 13-1), and mitigations such as effective environmental enrichment are needed.

It is important that horse types are well suited to the available environment. For instance, thick-coated native hill and moor ponies are more resilient and much more suited to cold, windy, upland environments than thin-coated thoroughbreds.

Severe discomfort is caused by more extreme failures in husbandry, such as tying horses exposed to the sun or leaving them in uncomfortable tack. This can occur in many different contexts and countries, from poorly run eventing yards to horses that work by transporting tourists (Figure 13-2).

Figure 13-1. Individual housing in a stable restricts important normal behaviours.

Figure 13-2. Horses used in the tourism industry can suffer welfare issues due to poorly fitting tack, insufficient hoof care, limited access to shade and water, and little opportunity for rest.

F3. Freedom from pain, injury and disease

Shortcomings in husbandry can predispose to non-infectious disease and accidental injury, which, in turn, can predispose to infectious diseases. Some highly infectious diseases can establish themselves even in fit and healthy animals. Good biosecurity can help prevent disease organisms from entering premises and limit the spread within.

Some common sources of pain, injuries and diseases are outlined below to explore their welfare impact; further details on the actual conditions are available in numerous excellent clinical texts.

Dental disorders

The feeding regime may not adequately wear down the continuously erupting, permanent, adult teeth. Diets with inadequate long fibre are a particular problem (see F1).

Captive horses generally need regular dental care from a qualified professional, such as a veterinarian or equine dentist, to identify and treat dental issues early – for instance, rasping sharp points. Behavioural signs of pain are subtle in horses and can be easily missed or misinterpreted. Even quite advanced dental pain could present simply as reluctance to be ridden, as applying tack (a 'bit') to the mouth becomes painful. Affected horses also risk impaction colic as they cannot properly bite and chew long fibre.

Metabolic disturbances

When horses are fed a diet high in grain and lacking in long fibre, excessive carbohydrates reach the hindgut and can upset the microbial balance. Metabolic disturbances are even greater in obese animals. Laminitis and/or colic commonly result, both acutely painful and sometimes requiring the animal to be euthanised.

Undernourished horses (see F1) can become underweight and predisposed to weaker hooves that crack and split more easily. They are also more likely to become ill by ingesting poisonous plants, as they lack alternative food sources.

Parasites

Internal and external parasites pose a serious disease threat globally. Risks vary with management practices. For example, where land is overgrazed, horses are more likely to ingest worm eggs as they are forced to eat close to their faeces, something they would normally avoid. Introducing infected animals to the group contaminates the pasture. Key mitigations include careful pasture management, routine screening of animals and strategic dosing with anthelmintics, whilst following guidelines to minimise anthelmintic resistance. Many horses are very irritated by flies and other insects, and effective preventative measures are needed.

Exercise types

'Free-living' horses engage in gentle to moderate exercise as they graze. This is beneficial for their foot health, as it sufficiently wears down their continually growing hooves and promotes good blood circulation in the lower limbs. The terrain underfoot varies greatly, from wetlands to rocky. Periods of more intense activity, such as running from predators, are relatively short. The exercise patterns of captive horses usually differ, the extent depending on their husbandry and usage. This can predispose to pain, injury and disease, and mitigations are needed.

In captivity, there are very different stressors on the feet and limbs, affecting hoof wear and health. Routine hoof care by suitably qualified people, including regular trimming and shoeing, is important. Details vary with factors such as age, living conditions and work type. Good hoof hygiene is an essential preventative. Horses need a clean, dry place to stand, whether in a stable or a field, and their feet should be regularly 'picked out'. Neglecting these leads to pain and injury, such as painful hoof cracks and altered limb-to-foot access. This places abnormal loads on the bones and joints of the lower limb, causing further pathologies, pain and lameness.

Captive horses are often trained and conditioned to gradually achieve the fitness level they need for the activities humans require of them. Even so, horses that are confined to stables for long periods and then suddenly exercised, and/or those subjected to very vigorous exercise, can be predisposed to injury and pain. Extreme examples include the injuries and occasional deaths suffered by horses competing at high speeds in endurance events (Figure 13-3), jumping large obstacles or being overloaded when transporting people or goods. Additionally, prolonged and intense exercise in hot and humid conditions can lead to physiological disturbances, resulting in hyperthermia, dehydration and electrolyte imbalances.

Tack and equipment

Captive horses often have equipment (tack) applied to their bodies for part or all of the day to perform the roles required of them. Ill-fitting equipment and the use of unsuitable materials can cause chafing of body parts and associated pain. Deliberate misuse of equipment, for example using whips to make animals race faster or work harder, causes fear and distress as well as pain and injury. Many other equipment-related practices have been little studied but could well cause pain and injury. Examples include using heavy chains or elastic bands attached between the horses' legs, or weighted shoes, to produce exaggerated limb action in dressage horses.

Mixing unfamiliar horses

Mixing unfamiliar horses upsets the social relations (see F4). This can lead to fighting, especially if horses are confined with limited opportunities to escape each other. Kick injuries are often worse if delivered by horses wearing metal shoes.

Figure 13-3. Horses used for racing undergo extreme physical exertion.

Mutilation (castration)

To avoid unwanted breeding in captivity, entire males and females cannot be kept together. Even where the sexes are kept separately, the behaviour of entire males may prove difficult to manage. Thus, the mutilation (see p.20) of surgical castration is commonly used to ease management, though there is some variation according to country and culture. This procedure inflicts pain, although the severity and duration can be mitigated by good practice in the provision of anaesthesia and analgesia. Both are typically administered in wealthier countries, such as the UK, where horses often have considerable emotional and economic value to their owners.

Level of veterinary care

Adequate veterinary care, both preventative and therapeutic, is essential for animal welfare. However, access to such care is often limited by cost, availability and owner awareness. Also, horses that have learnt to be fearful of veterinary professionals and veterinary interventions – for example, those that have become 'needle shy' (see p. 23) – can be difficult to handle and treat, making owners reluctant to seek veterinary help. In such cases, owners need guidance and support to desensitise their animals humanely and gradually.

Veterinary Primary Healthcare

Another key concern for veterinary professionals is ensuring they provide care that is appropriate for context. For example, a veterinary professional may have trained and developed their skills in wealthier countries but then apply these skills to equids in poorer countries, with different cultures, resources and opportunities. To achieve sustainable and meaningful improvements in welfare, standard procedures cannot be directly transferred because the same levels of technology and types of service are not usually practical, affordable and appropriate.

To optimise their effectiveness, veterinary professionals should first gain a thorough understanding of the new context, and often a more extensive skill set than is taught in traditional training in wealthier countries, as they face very different socio-economic, technical, and health challenges and opportunities to address these. Working with communities through participatory approaches and training is usually fundamental to achieving practical and ethical change. A primary health care approach is most likely to deliver the best welfare outcomes for the greatest number of equids, as it addresses the main problems within the local context and available resources, and strives for the most effective, socio-economically sustainable, evidence-based changes. This approach can be referred to as 'veterinary primary healthcare' and is often an integral part of a 'one health' framework for rural communities. 'One health' emphasizes that improving animal health and welfare also benefits human health and welfare. The welfare of equids is vital as many people, particularly in rural areas, rely on them for transport and work. Therefore, investing in animal welfare supports sustainable livelihoods, food supply and environmental sustainability. This, in turn, helps secure resources and encourages uptake at local, national and international levels.

Practical delivery involves multiple groups working together, including human health and other service providers, as well as veterinary services, engaging the whole community. It is important that animal owners and communities can take an active role.

Emphasising training and subsidiarity is essential. Training paraprofessionals can improve access and reduce costs for services, such as hoof treatments and abscess treatments. It is often more efficient to fund a few veterinarians to train and manage a multi-disciplinary team than to have all services performed by qualified veterinarians.

Primary healthcare aims to be economically sustainable in the long term. However, there is often an interim period in its development where local veterinary professionals require more significant and varied support than can be provided by the country or community itself. Viewing animal welfare as a 'public good' and a collective responsibility, rather than a burden borne by the poorest communities, makes the sharing of resources a logical progression. The global veterinary profession can enhance the welfare of working equines through knowledge sharing, advocacy and funding. WOAH has established international animal welfare standards for working equids, which are outlined in the Terrestrial Code[9] and can help inform national legislation, policy and implementation.

It is important that primary healthcare programmes provide veterinary treatments and do not focus solely on educating animal owners. Firstly, diseases and injuries cause significant suffering for both animals and people and, with modest resources, these can often be addressed for individual animals. Secondly, offering medical care builds trust within the community and fosters better communication, as having healthy animals is a common goal that most people want. Veterinary services should prioritise the main medical challenges, ensuring a minimum level of cost-effective quality service when and where required. Medical products for equids, previously too expensive for primary healthcare, are now more affordable due to changes in manufacturing. Certain analgesics, antibiotics and anthelmintics are often available at much lower prices in some lower-income countries than in developed nations. The challenge often lies in developing the supply chain and ensuring quality control.

Increased veterinary research on primary healthcare will support evidence-based treatments and offer alternatives to ineffective or harmful practices that are still prevalent globally.

F4. Freedom from fear and distress

Fear and distress often accompany pain, injury and disease. They also occur independently. Momentary fear and alarm are normal parts of everyday life, and unlikely to be of welfare concern provided they are infrequent and pass quickly with no adverse consequences for the horse. However, some situations can create more intense or prolonged fear and anxiety, which lowers their welfare.

Horses are prey species that have evolved to become easily fearful, quickly detecting and fleeing potential predators as a survival strategy. Despite this, they commonly learn to cope with a wide variety of situations they are routinely exposed to, which they would naturally be very fearful of. For example, they must manage being amongst crowds of people, the sights and sounds of attending equine competitions, and being regularly transported. They even learn to generalise this information, coping with new venues and environments relatively well. This is something people generally take for granted, but, on reflection, is quite remarkable given their natural tendencies. When horses have difficulty adjusting and fall short of human expectations, they are often unfairly labelled as having 'problem behaviours'.

To mitigate fear and distress, caregivers need to understand horse behaviour.

Human–horse interactions

People need to stay mindful of the general triggers for fear and anxiety in horses, anticipating and planning ahead in order to mitigate them. Some triggers can easily be overlooked, as things that often seem innocuous to humans can frighten horses. Their senses have evolved quite differently. For example, horses possess wide-angle, long-distance vision that is very sensitive to movement, evolved to scan the horizon for potential threats. They tend to be much more reactive than humans to sudden movements, such as flapping objects or people waving their arms. These can easily startle horses.

Unlike humans, scent informs much of the horse's worldview. They can move large volumes of air through their nose in one breath, trapping many scent molecules. They also have a special scent organ called the vomeronasal organ, which becomes more exposed to odours when they curl their upper lip (Flehmen response). Horses can use the odour of urine and faeces to identify other individuals in the herd, recognise the relative dominance of group members and identify their reproductive status. Unfamiliar scents, scents they have learnt to associate with unpleasant experiences (e.g. associated with veterinary interventions) and windy days that blow scents around can cause alarm. Loud noises can startle horses. Additionally, horses can detect and be unsettled by noises outside the human hearing range.

Horses vary in their propensity to fear according to factors such as breed, age, sex and individual 'personality' traits. Prior learning, whether from intentional training or from general life experiences, is a key determinant for how horses feel in different situations. They can adjust to many new situations and activities required by people if given sufficient opportunity to learn and properly socialised from a young age. Training should be humane, being based on making positive associations through reward and progressing at the right pace for that individual. Ideally, the horse will enjoy the process, and positive punishment should not be used (see pp. 20–24).

Horses are continuously learning, whether people intend to teach/train them or not. Previous unpleasant experiences can significantly impact their behaviour. For example, a horse that is usually calm around people may become alarmed by the presence of certain individuals, such as veterinary professionals, or when exposed to certain cues, such as the jugular vein being raised before venepuncture. They predict pain, and so react fearfully, usually by fleeing. If unable to escape, they are likely to defend themselves, for instance, by kicking and rearing. When this works for them, i.e. they gain immediate relief from the frightening situation, they are primed to repeat such actions. Since such fear-driven behaviours arise only in specific, infrequent situations rather than daily, they might go unaddressed. However, it is crucial for both horse welfare and human safety that horses are gradually desensitised to these situations to ensure they can receive appropriate levels of veterinary care.

Interactions with other horses

As horses are social animals, evolved to live in groups, being isolated from other horses often causes fear and anxiety. At times when they are physically separated, such as when individually housed in stables, they should at least be able to see, hear and smell other familiar horses.

Management choices can affect the quality of relationships between horses. Living in a group (herd) can cause fear if interactions with other group members become antagonistic (see F3), and they cannot escape one another. Unfriendly behaviours include chasing, biting and pinning back ears. These can be aggressive or used in defence, and tend to occur when horses compete for resources.

Certainly, care must be taken when mixing unfamiliar animals (see F3), and this should be avoided whenever possible. If done, introductions should be gradual and careful monitoring is needed as new hierarchies are established.

F5. Freedom to express normal behaviour

Normal behaviours

There are many breeds and types of horses, and their characteristics and behavioural tendencies vary. However, all horses share some innate, highly motivated behaviours, such as foraging and grazing, exploring, exercising, resting, living in groups, engaging in friendly social interactions with other horses, and performing body care activities such as grooming and rolling. They are highly motivated to synchronise many behaviours with others in the herd, a strategy that likely evolved to reduce predation risk. This synchrony is partly driven by 'social facilitation', where horses mimic the actions of nearby companions. It is also strongly driven by light patterns, with grazing and social activities peaking in the early morning and late afternoon, while resting mostly occurs during the middle of the day. Various factors, including temperature, season, forage availability and the presence of irritating insects, can also modify their behavioural patterns.

Providing horses with choices to perform these behaviours in a stimulating environment supports their welfare. Thwarting these behaviours can lead to unpleasant feelings such as prolonged boredom, frustration and fear.

In captive horses, the freedom to express normal behaviour varies with husbandry system and specific use. Well-managed, extensive group grazing most closely aligns with how horses live in their natural environment. They often retain considerable choice and behavioural freedom in a habitat for which they are well adapted, despite some restrictions on their movement, diet and social interactions. At the other end of the spectrum, some categories of horses tend to be kept individually housed and confined to a stable for extended time periods, and fed a high proportion of cereal-based feed and limited forage, and/or kept in tack and training/work for significant parts of the day. Here, normal behaviours are greatly restricted, and welfare risks can only be addressed with fundamental changes to the husbandry system type and horse usage. However, there are husbandry systems that sit between these two examples. Horses have a degree of adaptability, and further research could help identify when disruptions to their normal behaviours become problematic for them.

Many husbandry practices and, therefore, much of the current infrastructure for equine premises have developed according to traditions and habits rather than a good understanding of the horse's behavioural needs. For instance, horses typically have preferred companions, with whom they spend considerable time, often grooming each other and resting with their heads on one another's backs. However, it is often normal practice for leisure horses in European countries to be housed individually in stables (Figure 13-1) rather than in pairs or larger groups, despite the welfare benefits of this direct physical contact.

It can be difficult to change infrastructure due to factors such as awareness and financial cost. However, steps can be taken to optimise welfare within available premises. Environmental enrichment can benefit stabled horses by allowing them to engage in important behaviours, though not to their full extent. For example, to mimic foraging, horses can be given long forage, replenished at intervals to maintain their interest, and offered in various locations and containers where access presents a challenge. Studies indicate that many animal species display contrafreeloading – the need to express foraging and exploratory behaviours is important to them, not just the end result of obtaining food. They choose to 'work' for food through exploration and problem-solving, even when it is readily available.

Fixed brushes can be provided for horses to rub against, massaging their bodies, self-grooming and relieving itching. They can find interactions with people rewarding, such as being groomed or having their withers scratched.

However, it is challenging to provide enrichments that adequately cater for their exercise and social needs when horses are individually confined to stables.

Abnormal behaviours

Sometimes captive horses exhibit 'abnormal' behaviours, indicating a poor emotional state because their environment fails to meet their welfare needs. Examples include stereotypic behaviours such as 'box walking', 'crib biting' or 'windsucking'. Predisposing factors include confining horses to stables for extended periods, denying them access to sufficient dietary fibre and preventing them from interacting with other horses. It is important to correct the underlying deficiencies and avoid taking steps that merely stop the behaviour, such as applying a 'windsucking collar' or resecting neck muscles, to prevent horses from stretching their necks to windsuck.

Note that captive horses perform many behaviours that are not part of the 'normal' horse ethogram, but not considered 'abnormal' because they make sense in the context in which they are performed, being largely learnt by training from a young age, for example wearing tack and being ridden. Further welfare research could also provide insights into how horses experience the various activities they are used for, such as engaging in performance activities such as dressage, show jumping and racing, aiming to properly understand their preferences and motivations.

Welfare Assessment Protocol

AWIN (see Chapter 3) has developed a general welfare assessment protocol for horses[22] which, with training and further refinement, can be adapted for usage in different contexts.

The Brooke, a charity focused on the welfare of working equines in developing countries, has produced the Standardised Equine Based Welfare Assessment Tool (SEBWAT), which comprises 40 animal-based measures and is specifically tailored for the context of working horses in developing countries.[140]

Advocacy examples

- The British Veterinary Equine Association (BEVA) has produced reports on various practices that impact welfare, such as equine tack, embryo transfer, branding, use of the whip and racing pregnant mares.
- The Federation of European Equine Veterinarians (FEEVA) has a working group devoted to equine welfare and has produced multiple position statements on topics such as welfare in equestrian sport, during transport and euthanasia.
- Much horse husbandry is still based on tradition. Advocacy, even that undertaken by veterinary professionals, often focuses on improving

welfare within established beliefs rather than radically rethinking housing and management. However, a more holistic approach to welfare advocacy is emerging. A joint position paper by the FVE, FECAVA, WSAVA and the Federation of European Equine Veterinary Associations (2024) covers horses, dogs and cats. It includes recommendations for creating opportunities for behavioural expression and promoting appropriate social interactions, and discusses the relationship between animal behaviour, training and welfare.[114]

- Various veterinary associations support their members to be effective when called upon to act as witnesses in legal proceedings. For example, BEVA provides a 'Welfare case toolkit'.
- World Horse Welfare (WHW) is a UK-based charity which improves equine welfare through direct care, research, education, campaigning and collaborations with multiple stakeholders. They work globally to promote responsible involvement of horses in sport, serving as independent advisors to many governing bodies.
- Charities such as the Society for the Protection of Animals in North Africa (SPANA) and Brooke employ veterinary professionals who shape the approach to welfare and service provision for working equines in developing countries.
- Veterinary associations, such as the World Equine Veterinary Association (WEVA), span countries and cultures, aiming to improve welfare worldwide by offering quality continuing education for equine practitioners.
- In 2016, the World Organisation for Animal Health (WOAH) adopted recommendations for working equines.[9]
- Official guidance on providing for good welfare in horses has been published in the UK in the form of a Code of Practice under the Animal Welfare Act 2006.[141]
- In recent years, the EU has turned its focus to the welfare of equines kept for business purposes and is taking a similar approach to that used for farmed animal welfare. Significant resources are being allocated to review regulations, update them in accordance with scientific advancements, and develop guidance. Several years of advocacy led to the 2024 EFSA stakeholder meeting on equine protection. Key resources that preceded this included the 2018 EU report 'Guide to Good Animal Welfare Practice for Horses'[142] and the 'Removing the Blinkers'[143] report by WHW and Eurogroup for Animals, highlighting welfare challenges. So far, there are EFSA scientific opinions on equine welfare during transport[144,145] and slaughter.[145] The welfare implications of husbandry practices are being investigated.
- A European Reference Centre for Animal Welfare focusing on equines (EURCAW Equines and Ruminants)[146] has been set up. Its webpage will host various resources to support those implementing and enforcing welfare laws, and will also be helpful to anyone with a general interest in horse welfare.

- In 2025, the RSPCA published the report 'Horse Sense: A Good Life for Horses Through Strengthening the Horse–Human Relationship', which draws attention to the equine welfare crisis – driven by overpopulation, abandonment and neglect – and sets out recommendations for meaningful intervention.[147]

Fish

14

Abstract
The core welfare concepts discussed in Part 1 are applied to captive fish. Common welfare hazards are listed. Common welfare outcomes are identified and structured according to the Five Freedoms framework. Practical welfare assessment protocols are referenced. Examples are provided of how veterinary professionals have advocated for fish welfare or could be more involved.

Introduction

In this chapter, the term fish means 'finfish'. These are sometimes referred to as 'true fish' to distinguish them from other aquatic life whose common names end in 'fish', including molluscs such as cuttlefish, crustaceans such as crayfish, echinoderms such as starfish and other species such as jellyfish and shellfish.

Many fish share certain features. They are generally cold-blooded aquatic animals with gills that enable them to breathe in water and may have a swim bladder for buoyancy. They have fins for motion and stability. Most are prey species. Generally, they are oviparous, and they lay vast numbers of eggs to offset the high mortality rates of their offspring. Beyond this, fish are incredibly diverse creatures, with around 384,000 known species living in greatly different habitats and ecosystems worldwide. For instance, they are adapted for extreme temperature ranges, with some species living in very cold water (e.g. –2°C) and other in hot springs (e.g. 70°C). They may be pelagic (live near the water surface) or benthic (live near the bottom). They may live as a group (shoal or school) or be solitary. They inhabit fresh and saltwater environments.

Fish are the animal species most used by man. They serve various purposes, such as being used for food, as companion animals, for sport, in research and in public aquaria. Attention to fish welfare is a relatively recent development. The delayed interest may stem from people's difficulty empathising with fish

 Animal Welfare: A Concise Guide for Veterinary Professionals (K. Anzuino, J. Anzuino and P. Roger)
DOI: 10.1079/9781786391582.0014

compared with other animals. It may also be partly due to their general lack of knowledge about fish. For instance, people may presume fish live a natural life in their natural environment and are then quickly captured and killed.

Welfare research in specific fish species, particularly those newly domesticated, is sparse, which makes the practical application of the general welfare concepts challenging, especially regarding species' behavioural needs. Fish show evidence of sentience through their cognitive abilities, problem-solving skills, memory retention, pain receptors and stress responses. There is debate among scientists about the extent to which fish feel pain and emotions. For example, some argue that fish might not experience pain in the same way as mammals because they lack certain neural structures present in mammals. However, fish have evolved differently, and it is better to apply the precautionary principle.[148] After all, we do not conclude that fish cannot breathe because they lack mammalian-style lungs.

This chapter will illustrate general welfare concepts with examples from captive fish.

Husbandry

Most captive fish rely on human care, the extent depending on the husbandry system. For instance, highly controlled, artificial environments leave them totally dependent on humans, whereas other systems rely largely on natural ecosystems to support the fish. Most fish used for human food will be either commercially farmed or wild caught from natural habitats by commercial 'capture fisheries'. The commercial value of farmed fish now exceeds that of wild-caught fish.

Farmed fish

Around 580 fish species are farmed worldwide, with carp, tilapia, salmon and catfish being the most common. Fish farming ranges from subsistence to large-scale, intensive systems, and is the fastest-growing animal production industry. Intensive salmon farming in Norway, Scotland and Chile accounts for a small portion of total global fish production but yields the highest value per unit of production.

Key features of fish farming that enhance production compared with 'free-living' fish are enhanced feeding regimes, increased stocking densities and protection from predators. Farmed fish undergo various routine management procedures, including handling for grading, vaccinations, veterinary treatments, transportation and slaughter. Methods vary according to the farming system. For example, reproduction can range from that closely resembling spawning under natural conditions to highly artificial techniques, such as stripping eggs from broodfish under anaesthesia.

Farmed fish are produced using various systems, from ponds and sea inlets to highly intensive Recirculating Aquaculture Systems (RAS), depending on species and economics. Individual farms may use more than one system, changing according to the fish's developmental stage. Pond systems (Figure 14-1) are the least technical, often suitable for fish of lower commercial value, that are adapted to eutrophic environments. They consist of large earthen ponds with a natural food web, supplemented with additional feed to enhance production. Cage systems are floating frames with flexible netting used to hold fish in place, located in marine environments or lakes (Figure 14-2). Land-based

Figure 14-1. Farmed trout kept in ponds.

Figure 14-2. Large-scale salmon farming in Scotland seawater inlets (lochs).

flow-through systems are man-made basins or canals supplied with water from the surrounding landscape, such as wells or reservoirs. The water continuously moves through the system via specially designed inlets and outlets. Recirculating Aquaculture Systems (RAS) are highly technical land-based systems that focus on water conservation and biosecurity by recycling and purifying water for reuse. They are increasingly popular for the intensive farming of high-value species, such as Atlantic salmon.

Commercial capture fisheries

In capture fisheries, fish generally live in their natural habitat, usually a marine environment, until they are caught and retrieved. A range of capture techniques is used, the most common being trawling, purse-seine nets, gill or trammel nets, and pots. Generally, fish are crowded when caught in fishing gear and may be towed long distances before being retrieved from the water and landed on the deck of the fishing vessel.

Recreational fishing

In recreational fishing, often the fish is caught with a 'rod and reel'. This usually involves the fish biting onto a baited hook, being drawn to the water surface by tension on the line, then lifted out of the water using the line, and carried ashore (landed) using a net or a person's hands. The live fish may be photographed whilst lifted in the air. Fish are either returned to the water after hook removal or killed.

Fish as companion animals

Fish are found in various venues, including private homes, public aquaria and zoos. Their environments vary, from the highly controlled artificial environments of aquaria to more natural pond habitats. Approximately 1.5 billion pet fish are traded annually worldwide, with the majority being freshwater species. Most freshwater species are bred in captivity, while most marine species are wild-caught.

Laboratory fish

Fish are the second most common laboratory animals, after rodents. They are prominent, in part due to the increasing research aimed at supporting the rapidly growing aquaculture industry, and in part due to some species being excellent models for basic and applied research in fields such as physiology, genetics, ecology and behaviour.[148] Zebrafish have features that make them particularly important to the latter, such as producing transparent eggs and larvae that can be readily observed and manipulated, as well as sharing around 70% of their genes with humans.[148] They are small with a high reproductive rate, making them cost-effective to maintain.

Common welfare hazards

- Feeding regime.
- Water quality.
- Exposure to infectious agents.
- Stockperson to fish ratio.
- Difficulty observing animals.
- Handling and killing methods.
- Level of veterinary care.
- Lack of welfare and behavioural knowledge.

Welfare hazards are the resources (inputs) that can lead to poor welfare outcomes. The likelihood of poor welfare outcomes and their severity (i.e. the risks) can vary. Severity encompasses both duration and intensity. Common welfare outcomes are identified below, structured by the Five Freedoms framework.

Fish welfare is determined by how well fish can adapt, both physically and mentally, to their environment (see p.8). Relevant descriptions of 'free-living' fish are provided throughout to highlight the adjustments they must make while in human care. Here, 'free-living' is defined as fish living freely in an environment with the resources for which they have evolved and making their own choices. However, such comparisons provide only part of the welfare picture; welfare science (see Chapter 2) is essential for a proper understanding of the welfare impact of different husbandry systems.

Key determinants of welfare are the knowledge, skills and motivations of people caring for the fish, as well as whether they have sufficient resources (see pp. 24–25). Where these factors are optimal, welfare will be as good as it can be within a given husbandry system. However, there are often persistent welfare issues that are challenging to resolve without altering the husbandry system itself.

Common welfare outcomes

F1. Freedom from hunger

Thirst is generally considered irrelevant for fish, as they live immersed in water. However, they could potentially experience thirst if placed in water of different salinity or other physical/chemical content from that for which they are adapted. They are likely to experience hunger, though the sensation might differ from that in mammals, particularly in the fish species that do not possess a stomach. The quality, format, quantity and timing of feeding are likely important factors influencing sensations of hunger and satiety, as well as whether the fish is properly nourished.

The nutritional needs of different species vary greatly; for instance, they may be herbivores, carnivores, omnivores or detritivores. Commonly farmed, high-value species like Atlantic salmon have been well-researched, and properly formulated commercial diets can meet their nutritional requirements for production. However, these are the exception, and little is known about many other species, with sometimes only their natural history to refer to and extrapolate from.

Inappropriate diet

Inappropriate diets can lead to malnutrition and negatively impact health. For example, vitamin C deficiency can cause spinal deformities in catfish. Improper feed storage can reduce its quality and introduce potential pathogens. Hobbyists may buy large bags of fish food for cost savings, but leave these bags open in damp environments for many months. This lowers the nutritional quality. It also allows moulds to grow, exposing the fish to fungal infections and diseases such as aspergillosis. Fish are likely to become hungry and malnourished if the physical format of the feed does not suit their feeding behaviour. Some species or developmental stages may require more natural diets, such as bloodworms, krill, algae or whole dead fish, rather than standard food formats such as pellets, flakes or gels.

Environmental factors

Multiple environmental factors are likely to affect how well fish eat. For example, farmed Atlantic cod kept at low light intensities feed less, probably because the feed is less visible. Understanding the feeding patterns of fish and how environmental factors influence them can help us align food provisions with their natural appetite and satiety patterns. Where fish are exposed to the natural world, various annual, tidal and lunar cycles are likely to influence them in addition to more commonly considered factors such as temperature and light intensity. Much further research is needed. Excessive competition from other fish is another factor.

Routine food withholding

Food is often withheld before certain husbandry procedures that involve 'crowding' the fish in order to prevent a significant decline in water quality ('crowding' is the temporary gathering of fish in a smaller volume of water at a higher stocking density, done to facilitate handling). Withdrawing food before slaughter also improves food hygiene as there are fewer faeces in the gut to contaminate flesh during processing. Fish are likely to become hungry at times when food is completely withheld for several days. Withholding periods have generally been up to 3 calendar days. However, there is a move towards measuring withholding periods in 'degree days', which accounts for the effect of temperature on fish physiology. This refinement can minimise food withdrawal periods to what is strictly necessary by providing more accurate estimates for gut clearance time.

Under- and overfeeding

Fish may also be overfed, leading to obesity. For example, when different species of ornamental fish are kept in the same tank, some may be underfed, while others may be overfed, as it is challenging to cater to their diverse, species-specific needs when offering a single food source.

F2. Freedom from discomfort

Water quality

Good water quality is essential for good welfare – the consequences of poor water quality range from discomfort to illness and death. Fish rely on water for all bodily functions, including breathing, feeding, excreting, maintaining salt balance and reproduction.

Water quality concerns the water's chemical, physical and biological characteristics, including dissolved oxygen, carbon dioxide, ammonia, nitrite, nitrate, temperature, pH, alkalinity, water hardness, salinity and suspended solids. These parameters are interconnected, and the requirements of different species vary greatly.

Many captive fish are held in highly artificial environments, so keepers must recreate vital elements of their natural ecosystem. Dissolved oxygen, ammonia and temperature are particularly important, and discussed further below.

Fish need dissolved oxygen to breathe, while ammonia is their main waste product. Temperature strongly influences their metabolic rate.

Natural open bodies of water

In natural open bodies of water, there is enough oxygen to support the density and species of fish and ammonia levels are kept in check. Oxygen enters the water through direct diffusion at the air–water interface and through plant photosynthesis. This creates an oxygen curve where the dissolved oxygen level is lowest at dawn and highest in the evening. Levels are also significantly influenced by the stillness of the air–water interface and temperature; greater agitation increases dissolved oxygen, while higher temperatures decrease it.

The nitrogen cycle helps balance ammonia levels by converting it into nitrite and then nitrate, which plants, such as algae, use. Whilst nitrite is toxic to fish, nitrate is only harmful at high levels. Specific types of bacteria located in the surface layers of mud in ponds are responsible for converting ammonia to nitrate.

Highly artificial environments

In highly artificial environments, such as tanks containing ornamental fish or Recirculating Aquaculture Systems (RAS) for farmed fish, mechanical aeration is usually needed to ensure enough dissolved oxygen for breathing. Steps must be taken to prevent ammonia levels from rising. Elevated ammonia levels are likely to cause discomfort, whereas more extreme or chronically elevated levels can cause ill health, for instance, gill damage and an increased risk of bacterial infections and death. High levels are toxic. Ammonia levels are usually controlled by providing a suitable environment for the nitrifying bacteria to thrive, typically a properly maintained biological filter. This is usually unnecessary in less artificial systems (ponds, cages and flow-through systems) as ammonia is either naturally broken down or carried away by the water flow.

Fish factors

Fish factors such as stocking density and activity levels impact water quality – the greater they are, the more oxygen fish use and the more ammonia they excrete. Uneaten feed decomposing in the water is an important source of ammonia. This occurs even in fish farming, where waste feed carries a production cost.

Temperature

Fish are ectotherms, which means they take on the temperature of their environment. Temperature changes significantly alter their metabolic rates. Water temperatures outside the species' normal range will likely cause discomfort and predispose them to ill health. They also poorly tolerate rapid temperature changes. Large bodies of water, such as the sea and lakes, can hold large amounts of heat with relatively little temperature change, buffering fish against wide variations in ambient temperature. However, small volumes of water, such as those in small ponds and aquariums, cannot do this, placing fish at greater risk of large temperature fluctuations. This issue particularly affects fish farms that take in water from shallow rivers. For instance, temperatures in a trout farm can reach 27°C and above, while trout prefer to be at 17°C or lower. Careful husbandry is needed. For example, owners of ornamental fish tanks should measure and monitor the water temperature, take early remedial action to correct it, and be particularly careful when refreshing the water in the tank.

Salinity

Marine and freshwater species have different salinity requirements. In some species, the salinity needs can change significantly during their development. For example, wild Atlantic salmon hatch in fresh water, move to the ocean as young adults and then return to fresh water to reproduce. Appropriate salinity levels are important for fish welfare, but fish may also have a degree of adaptability. For example, in RAS, marine fish are often raised at lower salinity than in the wild for better production efficiency.

Gradual changes in salinity may be more important for welfare than specific levels, as long as they stay within comfortable limits.

Monitoring and managing water quality

Water quality parameters are colourless and odourless and can only be determined through measurement. Ongoing monitoring and adherence to recommended limits are essential. The 'worst-case' scenario should be considered. In fish farming, the lowest oxygen levels and highest ammonia levels tend to be found where fish congregate and where water flow is lower, such as at the centre of fish schools and the end of raceways or outlets.

Fish farmers generally understand the importance of water quality. Hobby keepers may be less aware and should be encouraged to purchase as large a tank as possible, as it is easier to maintain the quality of larger volumes of water. Changing a proportion of the water at regular intervals and measuring water quality parameters weekly is advisable.

Other environmental factors

Environmental factors other than water quality are likely important for fish comfort, but our understanding is limited. For example, lighting patterns and shelter provision could determine whether fish receive sufficient rest.

F3. Freedom from pain, injury and disease

Sources of pain, injury and disease

Non-infectious diseases generally result from suboptimal husbandry, such as inadequate nutrition and poor water quality. Fish rely on water for all their bodily functions, so water quality is vital for health (see F2).

Physical injuries often result from trauma and are likely to be painful. Examples include the fin bite injuries experienced by fish attacked during outbreaks of aggression, and injuries that occur when fish collide with enclosure structures and other fish, particularly during crowding or intense schooling at high stocking densities (see F5). These factors immunosuppress fish and increase susceptibility to infectious pathogens. Fish are constantly bathed in water containing potential infectious pathogens, another important component of water quality. Good biosecurity and the usage of sterilisation technologies such as ozonation reduce the number of pathogens, but do not eliminate them.

Levels of veterinary care

Fish may not receive an appropriate level of veterinary preventative interventions and treatments. Awareness among fish owners may be limited in some sectors. Veterinary provision may be seen as less important than in terrestrial animals, partly due to less interest from veterinary professionals, as well as less training, and partly due to owners being less emotionally attached to their fish than to other companion animal species. In contrast, larger commercial fish farms with high-value fish are more likely to use health and welfare planning in conjunction with a veterinary fish specialist.

Refinements

There are multiple areas where welfare and production can improve in tandem, thereby giving impetus for change and meeting the welfare potential of the system. Examples are early and accurate disease detection through routine screening, refining the environment to remove sharp edges or abrasive surfaces, using appropriate covers on the water inlet and outlet pipes to prevent fish entrapment, and minimising the time fish are out of water during handling and transport.

Further refinements include using fish pumps to transport fish to the location where they will undergo husbandry procedures or slaughter, which keeps them in water until the last possible moment. Fish grading equipment can be refined by using specialised grids to allow specific sizes of fish to swim through while preventing others from doing so, instead of relying on manual handling for sorting different fish sizes. Many farms use mechanical graders, which are not quite as welfare-friendly as passive graders during the actual handling process. However, there are welfare 'trade-offs'. Mechanical graders allow for superior grading, which provides longer-term welfare benefits. For example, there will be less size disparity and therefore, less competition for food following mechanical grading.

Early and accurate disease detection

Early detection of disease and injury is challenging. Selecting a representative sample to screen can be challenging when working with extremely large fish numbers in sectors such as fish farming. Also, fish may be difficult to observe due to their underwater environment, making it difficult to monitor their feeding and other behaviours. Some farming systems benefit from the use of underwater cameras, which provide better visibility than simply observing the water surface.

'Cleaner' fish

High-value farmed salmon are often raised alongside wrasse or lumpfish, which are referred to as 'cleaner' fish. These cleaner fish feed on sea lice, providing a natural alternative to medicine usage for controlling sea lice infestations in salmon. The welfare of the cleaner fish should be considered of equal importance to that of the salmon, not overlooked due to their lower economic value.

Interpreting fish behaviours

Understanding and interpreting fish behaviours can be challenging, and much further research is needed.

Signs of pain in fish are relatively unknown, and the limited evidence suggests that they vary by species. For instance, rainbow trout and zebrafish show increased ventilation rates in response to nociceptor stimulation, but carp do not. Zebrafish reduce their swimming rate when nociceptors are stimulated, whereas rainbow trout increase theirs. Injection of acetic acid into the lip elicits 'lip rubbing' in rainbow trout, thought analogous to humans rubbing painful areas. In zebrafish, acetic acid injections near the tailfin are linked to rapid fin movements (tailfin fanning).

Pain recognition is especially relevant for laboratory fish because they are intentionally and routinely subjected to invasive procedures. In the UK and EU countries, fish must be anaesthetised for most invasive procedures. Further studies are needed on the effectiveness of anaesthetics and analgesics. Again, different species present different challenges. For instance, determining muscle relaxation and analgesia in very small species such as zebrafish is challenging.

Vaccine and medicine availability

Other factors likely to limit veterinary interventions include a dearth of authorised, effective, preventative medicines, such as vaccines that are practical to administer, and very few licensed treatment options. In fish farming, there is also the practicality of administering medication to very large numbers of fish. Treatment regimes are often stressful for fish, requiring them to be crowded, handled and immersed in various chemicals or drugs.

Extreme conformation

Some companion (ornamental) fish are bred for extreme conformation, prioritising appearance over physical fitness and function. For example, Oranda goldfish may suffer impaired vision due to excessive growth of the head crown ('wen'), telescope goldfish have protruding eyes that are predisposed to trauma, and rotund-bodied goldfish have impaired swimming ability and are prone to swim bladder problems.

Capture and killing methods

Capture and killing methods impact welfare. In some settings, such as large commercial fish farms and commercial capture fisheries, vast numbers of fish must be managed simultaneously.

In fish farming, there is more control over the process compared with commercial capture fisheries. Better handling, stunning and killing methods are being developed for high-value, intensively farmed fish species, such as Atlantic salmon and trout. Unless the animal is made immediately unconscious and insensible to pain, killing methods can cause pain, distress and suffering. Welfare issues are often linked to transport and handling, which are necessary prerequisites for killing fish (see Chapter 17).

Commercial capture fisheries pose differing welfare problems to fish farming that are logistically very difficult to address. Issues have been identified, but it is presently unclear what the solutions are. Fish are free-living in their natural habitat until captured. All capture techniques cause varying degrees of pain, injury, fear and distress. Generally, fish are crowded when caught in the fishing gear and may be towed long distances before retrieval. They may be dragged and compressed for hours in trawl nets. There is ongoing welfare research focused on improving net design and fishing methods, including using electrical stunning with purse seine nets in commercial fishing.[149] Fish suffer fatigue, exhaustion and physical injuries from excessive strenuous escape attempts inside the fishing gear and when landed. They suffer low oxygen levels when crowded in nets, progressing to physical constriction when packed amongst other fish, making breathing difficult. Many are still alive when retrieved from water and experience being rapidly lifted from the water depths and into the air. As they ascend, the rapid drop in water pressure causes their swim bladders to expand or burst and gas bubbles form in their blood and organs. These factors, along with the sudden temperature change, massively disrupt their metabolism. Outside of the water, they cannot breathe at all. They also lose crucial body support provided by water pressure and are exposed to much greater light intensities than they experience at normal swimming depths. They then typically die from suffocation, freezing on ice, or being eviscerated, and it takes them considerable time to lose consciousness. Another issue in capture fisheries is 'ghost fishing', where lost fishing gear remains in the ocean and continues to entrap fish for extended periods. Such fish are never retrieved and slowly die.

F4. Freedom from fear and distress

Applying the precautionary principle, we can assume that fish can feel unpleasant emotions unless proven otherwise. Feelings akin to fear and distress will likely accompany pain, injury and disease, but they could also occur independently.

Fish are likely to experience fear when held captive near predators, whether the predation is real or perceived. Similar to mammals and birds, they have protective responses to avoid predators and well-developed physiological stress responses.

Captive fish may be preyed upon. Ornamental fish owners with insufficient knowledge may place predator and prey species in the same tank, and also provide inadequate shelter, making it difficult for prey species to escape. In fish farming, it can be challenging to prevent attacks. Many fish farming systems are situated in remote, rural areas, partly open to the natural environment, where the abundance of confined fish attracts predators, commonly sharks, otters, herons, seals and sea lions. Deterrence is challenging; for example, seals are known to jump over high gates and suck fish through nets. Using acoustic technology is an option, but it can inadvertently harm fish and other wild animals. It is now banned in the UK to protect marine mammal welfare.

Fish may sense a predation risk, even when they are not under direct threat of attack. Signals include the scents released by human skin, the sight of people and objects looming over the water surface, and noise and vibrations from equipment, such as boats and pumps. Crowding, confinement, sudden environmental changes and disrupted social groupings are all threatening. Fearful, stressed fish swim more rapidly and make escape attempts, which can cause physical injury.

F5. Freedom to express normal behaviour

This section builds on discussions of fish behaviour under the other Freedoms. Our understanding of the connection between welfare and behaviour in fish is limited. Ethograms are unavailable or incomplete for many species, and scientific techniques for investigating behavioural priorities, motivations and preferences are underdeveloped. For instance, we are unsure whether Atlantic salmon undertake long migrations primarily to seek food or if they have an innate drive to travel specific distances and directions, as in migrating birds.

It is possible to extrapolate from first principles in other species, creating hypotheses, but caution is also needed. For example, reduced stocking density is often recommended to improve the welfare of many terrestrial farmed animals, but fish may differ. The optimal stocking density may vary by species and situation. Low stocking densities may lead to increased aggression due to territorial behaviour, while higher stocking densities can prevent the fish from establishing territories in the first place, reducing aggression. There may be an optimal lower limit below which welfare will not improve and, indeed, may worsen.

Another interesting research finding is that some fish have been shown to move to warmer waters when they are ill, which improves their survival rates and is thought to be analogous to the fever mechanism in warm-blooded animals.

Despite the overall dearth of information, basic knowledge of fish behaviour patterns is being used practically to improve welfare and production in some contexts. For example, underwater camera technology in some salmon farms has proved useful for monitoring group activity and identifying whether the fish are underactive, overly active or exhibiting signs of aggression. Findings can help farmers match feed delivery to fish appetite, reducing food waste and its associated costs, as well as detrimental effects on water quality. This also limits aggression. Knowledge about how factors like temperature, light and time of day can influence fish behaviour and movements can potentially be used to improve welfare during handling.

Environmental enrichment

The artificial environments where captive fish are kept often lack adequate stimulation or enrichment. Most intensive fish farms are barren. Where enrichments are attempted, such as in ornamental fish tanks, they tend to be based on tradition and practicality rather than a solid scientific understanding of fish welfare, behaviour or even the species' natural history.

However, dietary, physical and social factors could be changed to allow for more complete behavioural expression and provide mental challenges that promote welfare.[150,151] The limited research primarily focuses on laboratory zebrafish and suggests that stimulating environments are beneficial as they enhance the fish's cognitive development, brain size, spatial mapping skills and memory. For example, juvenile zebrafish raised with plants showed better brain development than those in barren environments.

Food types and presentations could be made more interesting. Fish in natural-style ponds probably already live in an environment that caters well to many aspects of their welfare, having a diverse diet and some natural prey. However, those in artificial environments are often fed formulated foods without the opportunity to be selective. There is a lack of knowledge on how to make feed provision more enriching for different fish species. Consider a marine species, such as sea bream. They naturally feed on hard-bodied prey that they must chew, and they tend to be very wasteful when fed standard pellets, as they manipulate and break them into small pieces. However, we do not understand their motivations – whether they have a strong behavioural need to chew hard foods or whether they are simply 'playing' with these pellets.

Providing substrates, like gravel, on the tank floor can potentially improve foraging behaviour. Shelters can provide comfort and an escape from other tank inhabitants. However, research is needed to identify design details that make these meaningful for the fish.

Other environmental factors, such as colour and lighting patterns, will likely impact behaviour and welfare. For example, the tank background colour appears to affect turbot growth. Simulating natural dusk and dawn lighting rather than subjecting fish to sudden on-off lighting will reduce stress.

Promoting positive social interactions could also benefit welfare. However, catering to the diverse behavioural needs of mixed-species shoals can be challenging. It is difficult to replicate their natural ecosystem, and certain aspects, such as predator–prey interactions, are undesirable for ethical reasons. Environmental conditions likely influence social harmony within a fish shoal. Some general principles to consider include providing enough shelters or hiding spots for the number of fish, ensuring that prey species have hiding spots and shelters that only they can access to evade potential predators and providing enough resources for subordinate fish to escape from dominant ones. The sex of the fish should be considered, as sexual and territorial aggression within a species can significantly impact behaviours.

Welfare assessment protocols

Welfare assessment protocols for captive fish are being developed using concepts similar to those for terrestrial animals. More research is needed to identify effective resource-based and animal-based measures, especially for evaluating large numbers of fish underwater. Resource-based measures include water quality, nutrition and habitat design. Outcomes vary by species and husbandry system but are likely to include body condition, fin condition, colouration, growth rates, food intake and swimming behaviour during feeding. Fish are generally more difficult to observe than most terrestrial animals. Measures like 'crowd scoring' assess fish activity levels by rating water surface movement on a scale of 1–5, where 1 indicates little movement and 5 indicates intense splashing. Underwater cameras can also improve fish visibility.

Currently, protocols are more advanced for farmed and laboratory fish than for companion (ornamental) fish, which have seen little development.

Farmed fish

Protocols have been most extensively developed for commonly farmed, high-value species such as Atlantic salmon and rainbow trout. Examples are the FISHWELL handbooks, created by fish welfare researchers and veterinarians from Norway and the UK and hosted by Nofima, providing welfare indices and assessment protocols for Atlantic salmon, rainbow trout, gilthead sea bream and European sea bass.[152]

APROMAR, the association of Spanish fish producers, has released guides to improve welfare practices in fish farming. The first volume, published in 2022, addresses key welfare concepts for fish.[153] Guides for European sea bass and gilthead sea bream followed in 2024, with further guides planned for other species and production systems.

FAI has developed a free, easy-to-use Welfare Assessment App for Tilapia, supported by educational materials.[153]

The European Reference Centre for Animal Welfare for Aquatic Animals (EURCAW Aquatic Animals),[154] created in 2024, will be developing and disseminating information on welfare indicators, welfare assessment methods, and tools to improve the welfare status of aquatic organisms.

Laboratory fish

A greater range of behavioural measures can be practically assessed in contexts where a smaller number of fish are easily observed, such as in laboratory settings or specific aquaria. Several general behaviours indicate distress in fish, including:[155]

- Fin clamping.
- Colour changes, such as darker hues indicating stress or paler colours due to excess mucus production when the fish are parasitised.
- 'Flashing', which is when a fish rubs against substrates or objects.
- Gasping, especially at the water's surface.
- An increased opercular beat rate, which suggests a higher respiratory rate.
- Buoyancy issues.

Methods for observing and monitoring behaviours are being developed, such as the Fish Behaviour Index (FBI).[156] This automated tool monitors the behaviour of singly housed zebrafish, a common laboratory animal, and can detect subtle behavioural changes that may indicate pain.

To date, there has been little progress in developing welfare measures and assessment protocols for companion (ornamental) fish.

Advocacy examples

- Organisations with specific expertise in animal welfare are increasingly focused on improving fish welfare:
    - The RSPCA has developed detailed welfare standards for farmed Atlantic salmon and farmed rainbow trout.[157,158]
    - FAI has created free online courses to educate people about fish welfare and welfare assessment, as well as to support the use of the Tilapia Welfare Assessment App that they developed.[153]
    - The NC3Rs is developing and disseminating information on the practical application of the 3Rs in laboratory fish.
- Various industry bodies and scientific associations provide a wealth of information about fish in different contexts. Examples include The Fish Site, which focuses on aquaculture,[159] and organisations such as the Ornamental Aquatic Trade Association (OATA) and Ornamental Fish International (OFI), which provide insights into ornamental fish. The World Organisation for Animal Health (WOAH) has released a free mobile app called 'Aquatic Disease Field Guide', which provides easy access to information on aquatic animal diseases as outlined in the WOAH Aquatic Animal Health Code. The app includes diagrams and photographs of anatomy. Recently, various stakeholders have begun to focus specifically on fish welfare, embracing and learning to apply the welfare concepts.

- Currently, most assurance schemes for fish products address social and environmental issues, ensuring that fish are sustainably caught or farmed. Very few assure welfare. Notable exceptions are the RSPCA Assured schemes for Atlantic salmon and rainbow trout in the UK. The development of more robust schemes for assuring environmental protection standards is crucial for the welfare of wild fish and many other wild animals. For example, monitoring and regulation have struggled to keep pace with the growth of the fish farming industry. Pollution is a significant concern in fish farming with open-net cages, as contaminants easily enter surrounding waters in the form of uneaten feed, metabolic waste, chemicals such as antibiotics and disinfectants, and disease-causing organisms. Additionally, farmed fish often have less genetic diversity than their wild counterparts. When they escape, through accidental releases or predator attacks on infrastructure, they may breed with and threaten the resilience of wild populations.[160]
- Veterinary professional associations are increasingly advocating for fish:
  - The BVA policy position on sustainable finfish aquaculture in the UK (2023) emphasises fish welfare while acknowledging the importance of environmental and human health.
  - The Fish Veterinary Society (FVS), a specialist division of the BVA, informs policies and consultation responses, such as the Animal Welfare Committee's call for evidence regarding the killing of farmed fish.
  - The World Aquatic Veterinary Medical Association (WAVMA) expressed support for the Fish Welfare Dialogue Webinar (2024), which introduced and discussed the challenges faced by those working in fish welfare.
- Veterinary professional associations can raise students' awareness about career opportunities in fish veterinary medicine and advocate for its inclusion in veterinary training, as this area is often overlooked despite the vast numbers of fish used by humans.
- In some countries, there is a growing interest and lobbying to grant fish legal welfare protections similar to those of terrestrial farmed animals:
  - The Scottish Parliament has discussed differences in legal protections for farmed fish and land animals during a meeting of representatives from the aquaculture industry and animal protection organisations in early 2025.
  - The EU has adopted fish welfare as a serious area of concern that requires better mandatory regulation.
- WOAH has recognised that fish have a welfare state and that recommendations are needed to protect their welfare. Currently, the welfare section of the Aquatic Animal Health Code (2024) is brief and general, but is expected to develop further.[9]

- The UK Farm Animal Welfare Council report (2014), updated by the recent Animal Welfare Committee report (2024), focuses on farmed fish welfare, with particular emphasis on welfare during transport and killing.[161]
- Many EU resources are posted on the EFSA fish welfare pages. The report titled 'Animal Welfare of Farmed Fish', published by the European Parliament, is a particularly useful resource that covers species of fish commonly farmed in the EU.[162]
- EURCAW Aquatic Animals, created in 2024,[154] will post publications, news and training materials on its website.
- Some organisations are trying to make better use of existing regulatory tools, such as licensing. An example is the work of OATA, aiming to improve the licensing of pet shops.

Wild Animals

15

Abstract

The core welfare concepts discussed in Part 1 are applied to wild animals. Common welfare hazards are listed. Common welfare outcomes are identified and structured according to the Five Freedoms framework. Examples are provided of how veterinary professionals have advocated for wild animal welfare or could be more involved.

Introduction

In this chapter, 'wild animals' are defined as undomesticated animal species, naturally found free-living in the wild. This encompasses vast numbers of highly diverse animals. Accurate figures are unavailable, but estimates suggest that global wild animal numbers far surpass those of domesticated animals and humans by several orders of magnitude.[163] There are an estimated 100–400 million birds, 100 billion–1 trillion mammals, 100 billion–100 trillion reptiles, 100 billion–100 trillion amphibians, and somewhere between 10 trillion and over 1000 trillion fish alive at any one time, compared with only 24 billion terrestrial vertebrate livestock and 9 billion humans.[164,165]

Human actions directly or indirectly impact most free-living wild animals. In addition, wild animals are held captive by humans, temporarily or permanently, for various reasons.

Wildlife–human interactions

Human interactions with wild animals can be broadly categorised as follows:

- Free-living wild animals that are deliberately targeted.
- Free-living wild animals that are incidentally harmed.
- Captive wild animals.

DOI: 10.1079/9781786391582.0015

Free-living, deliberately targeted wild animals

Humans deliberately target certain free-living wild animals in ways that inadvertently cause them harm. Examples include population control strategies, often referred to as 'pest' control, and sports such as recreational fishing/angling (Figure 15-1).

Free-living, incidentally harmed wild animals

Humans can accidentally and often unknowingly harm free-living wild animals. Examples are the routine, everyday activities that support human lifestyles, including certain agricultural practices, urban development and transportation systems.

Attitudes to free-living wild animals

All in all, people have very mixed attitudes to free-living wild animals, shaped by culture. Some species tend to be automatically considered 'high value' and worthy of protection. In contrast, others tend to be seen as 'low value'. For instance, in the UK, many people are repulsed by wild rats, even though they are simply another sentient creature attempting to thrive.

When wild animals threaten human interests, posing disease threats or competing for food, they are labelled rhetorically as 'pests' or 'vermin', and may be subject to inhumane killing methods that would not be tolerated for domestic species of similar sentience. To illustrate, examples of traps used to capture wild animals in the UK are shown in Table 15-1. Lethal traps are

Figure 15-1. Recreational fishing (see Chapter 14).

designed with the intent of killing animals instantly, whereas live traps capture and hold them until they are killed by the operator or, in some cases, released.[166,167]

Table 15-1. Examples of traps used to capture wild animals in the UK[166,167]

Trap type		How it works	Main species targeted in the UK
Lethal	Spring loaded	A spring mechanism is triggered when the animal enters the trap, causing rapid force to be applied to the animal's body (Figure 15-2).	Stoats, weasels, mink, grey squirrels, rats, mice and rabbits.
Live capture	Live bird traps, e.g. Larsen (Figure 15-3)	Bait or decoy birds attract the target birds, which, once inside, are unable to fly out.	Magpies and crows.
	Snares	A wire noose fixed to the ground catches a running animal by the neck.	Foxes, rabbits, and brown hares.
	Glue traps	Strong adhesives applied to a surface firmly hold any animal part that comes into contact with it (see Figure 4-1 right, p. 56).	Rats and mice.
	Box and cage traps	Basic box with solid or mesh sides. The animal enters, attracted by the bait. Once inside, the door closes and locks behind it.	Foxes, badgers, mink, rabbits, and grey squirrels.

Figure 15-2. A mole trap is placed in a mole tunnel. The spring mechanism is activated when the mole passes through, trapping it in metal jaws.

Figure 15-3. A Larsen trap containing a magpie to attract and trap territorial rivals, which are subsequently killed. There are welfare issues related to the needs of decoy birds used as well as the trapped birds, including how they are killed.

Human preferences are not fixed. Species labelled 'pests' at one time and place are valued and protected at others.[168] Many free-living wild animals go largely unnoticed, and humans are often oblivious to the welfare harms their everyday actions cause.

Captive wild animals

Humans may hold wild animals captive, either permanently or temporarily, in what is collectively referred to as captive wild animal facilities. Examples include zoos, rescue rehabilitation facilities, sanctuaries and entertainment facilities like circuses.

Zoos traditionally were menageries solely for public entertainment. In some countries, zoos still fulfil this role. The narrative of many modern zoos is one of preserving biodiversity through conservation, research and educating the general public. In some countries, such as the UK and those in the EU, there is a legal requirement to do so. Sanctuaries may permanently hold and care for animals that cannot cope in the wild. Rescue-rehabilitation facilities, such as wildlife hospitals, provide treatments for injured or sick wildlife and temporarily support them until they are judged able to cope in the wild. In some countries, circuses can keep wild animals solely for entertainment, often requiring them to perform tricks for audiences. Wild animals are sometimes kept as companion animals (pets).

This chapter illustrates core welfare concepts using examples of human activities affecting wild animals, both living in the wild and captive in zoos, rescue-rehabilitation centres and sanctuaries.

Captive wild animals in other contexts are outside the scope of this chapter, but their welfare can be analysed similarly from first principles.

Animal welfare science

The welfare of wild animals has generally been less studied than that of many domesticated animals. In recent years, research into the welfare of captive wild animals, particularly those in zoos, has been rapidly expanding.

The welfare of free-living wild animals appears to be becoming higher profile. Wild animals deliberately targeted have received some research attention, in particular, the humaneness of different killing methods. Situations where animals are incidentally harmed have been far less researched, despite there being few environments globally unaffected by human activities,

In the absence of data for many free-living wild animals, metrics such as the rate of decline in their numbers have been taken as proxy indicators of welfare, assuming that it is preceded by morbidity, mortality and associated suffering. The rate of destruction of habitats on which animals depend is also a guide.

Before proceeding, it is worth considering the likely welfare of those wild animals that still live largely in an undamaged natural habitat for which they have evolved. Certainly, the wild should not be romanticised – animals will likely experience periods of hardship and poor welfare, for instance due to extreme weather conditions, food shortages and predation, something humans can protect them from in captivity. However, they will also likely experience significant periods of good welfare, living in an environment for which they are well-adapted, using their physical and cognitive abilities to the full and having considerable choice, something difficult to replicate fully in captivity.

Relationship between welfare and conservation

Sometimes, conservation and welfare are conflated. However, they are distinct disciplines. Traditionally, wildlife studies have been the domain of conservation. '*Conservation ... works at the levels of population, ecological systems, and genetic types, and deals with threats to biodiversity and ecological integrity. Animal welfare works at the level of individuals and groups and deals with threats to the animal's health and quality of life*' (Fraser, 2010).[168] Sometimes, welfare and conservation are in tandem. For example, animals suffer and die when ecological systems are disturbed. Other times, conservation and welfare are in conflict. For instance, UK conservationists expend considerable effort protecting red squirrel populations by suppressing grey squirrel populations. However, grey squirrels have become more successful due to their better adaptation to their environment, which is likely to lead to better welfare. Welfare and conservationists can benefit from better collaboration. For example, conservation researchers can contribute their skills in tracking and monitoring free-living wild animals to welfare scientists. In turn, welfare scientists could contribute their expertise in identifying and alleviating pain, thereby improving the quality of studies where animals have undergone painful marking procedures. Many citizens care about both welfare and conservation, so policies that address both are more likely to be successful.[168]

Common welfare hazards

The examples given below are by no means exhaustive.

Free-living, deliberately targeted wild animals

- Methods of pursuit, capture and killing.

Free-living, incidentally harmed wild animals

- Transportation systems.
- Agricultural practices.
- Waste disposal and pollution.
- Non-native predator species.

Captive wild animals

- Lack of domestication.
- Level of species-specific knowledge.
- Enclosure design.
- Level of environmental enrichment.
- Population management strategies.
- Geriatric animals.
- Euthanasia decisions and methods available.
- Hand rearing.
- Triage.
- Fitness for release assessments.

Welfare hazards are the resources (inputs) that can lead to poor welfare outcomes. The likelihood of poor welfare outcomes and their severity (i.e. the risks) can vary. Severity encompasses both duration and intensity. Common welfare outcomes are identified below, structured by the Five Freedoms framework.

Wild animal welfare is determined by how well wild animals can adapt, both physically and mentally, to their environment (see p.8). Relevant descriptions of 'free-living' wild animals are provided throughout to highlight the adjustments they must make while in human care. Here, 'free-living' is defined as wild animals living freely in an environment with the resources for which they have evolved and making their own choices. However, such comparisons provide only part of the welfare picture; welfare science (see Chapter 2) is essential for a proper understanding of the welfare impact of different husbandry systems and situations.

Key determinants of welfare are the knowledge, skills and motivations of people caring for or interacting with the animals, as well as whether they have sufficient resources (see pp. 24–25). Where these factors are optimal, animal welfare will be as good as it can be within a given husbandry system or during a given activity. However, there are often persistent welfare issues that are challenging to resolve without altering the husbandry system or activity itself.

Common welfare outcomes

F1. Freedom from hunger and thirst

All wild animals need access to sufficient water and food of the correct type and quality for their species and developmental stage. Otherwise, they will likely experience prolonged thirst, hunger and malnutrition.

Free-living, deliberately targeted wild animals

Trapped animals will often experience hunger and thirst. The extent varies according to the trapping method (Table 15-1). There is scope to provide some food and water within box or cage traps, though whether it is eaten depends on its suitability for the species and their levels of distress. Other live capture traps, such as snares and glue traps, prevent eating and drinking altogether. The duration depends on how frequently the traps are inspected and whether the operator kills the animals immediately. In some cases, it will be for many hours, and animals may be left to die. Young animals, dependent on their parents for food and care, are likely to starve if their parents are trapped. In recreational fishing, captured fish suffer painful mouth injuries from the baited hook and may find eating difficult once returned to the water.

Free-living, incidentally harmed wild animals

The quantity and quality of food and water available are reduced to varying extents where human activities denude and pollute habitats, impairing ecosystems.

Captive wild animals

Providing nutritionally balanced diets for different animal species can be challenging due to limited species-specific knowledge and the need to cater for multiple species. Monitoring can be challenging. For example, signs of nutritional deficiencies may be subtle, and assessing body condition can be difficult without appropriate scoring systems and regular handling for palpation.

F2. Freedom from discomfort

Free-living, deliberately targeted wild animals

The capture method determines comfort levels. Trapped live animals may endure inclement weather conditions, the extent of which depends on the trap site selection. Their ability to engage in basic comfort behaviours is limited. Whilst box or cage systems allow some trapped animals to lie and rest, and some even provide bedding, snares and glue traps do not allow for any comfort. In recreational fishing, landed fish may be kept temporarily in small basins of water before being released or killed. Water quality will deteriorate rapidly in these small volumes, especially when shared among several fish.

Free-living, incidentally harmed wild animals

Damage to natural habitats has consequences for animal comfort. For instance, the loss of vegetation covering the ground means less shelter from inclement weather and fewer resources for essential activities, such as den or nest building.

Captive wild animals

Enclosures should be designed to provide a comfortable resting environment, with species-appropriate resources such as bedding, perching areas and sufficient space for all animals in a group. Proper lighting, temperature and humidity levels are essential. Certain wild species may have unique needs, such as requiring ultraviolet light. Some may need a highly controlled, 'closed' environment because their needs differ significantly from those of the natural environment in the country where the captive facility is located.

F3. Freedom from pain, injury and disease

Free-living, deliberately targeted wild animals

All methods of pursuit, capture and killing cause some harm. The extent depends on the method and the operator's competence.

Trapping Spring-loaded lethal traps may fail to work as intended, causing serious injury and prolonged suffering rather than killing the animals outright. Animals in live capture traps often become debilitated and injured by making repeated escape attempts, even in the more humane box and cage traps. They risk attack from animals trapped alongside them. Snares and glue traps appear the least humane. Trapped animals suffer exhaustion and serious external and internal injuries as they attempt to escape, including those from self-mutilation. For example, those in snares suffer serious mouth injuries and may asphyxiate due to strangulation. Glue traps commonly lead to torn skin and broken bones, and struggling animals become further entrapped, with the eyes, mouth and nostrils becoming glued shut. Non-target species will also be captured and harmed.

Where distressed and debilitated live animals are released rather than killed, they may have difficulty coping in the wild, thus suffering further.

Poisoning Rodenticides, typically anticoagulants, are used to control wild mouse and rat populations by being placed in bait. The animals suffer a slow and painful death, usually over several days, after consuming a lethal dose.

Recreational Angling Fishing with a rod and reel inevitably injures fish. The baited hook becomes embedded in their lips, mouth, oesophagus or stomach. Their pain and injury are exacerbated when they are retrieved by pulling the attached line. Vigorous escape attempts leave them exhausted and debilitated. They become even more vulnerable to injury when in the air, as their body is no longer supported by water pressure, and they cannot breathe. Recreational anglers have the scope to pay attention to individual fish, something virtually impossible in commercial, capture fisheries (see p. 220), and can refine their practices to lessen harm. Examples include using less abrasive landing nets and distributing the fish's weight across both hands when lifted out of the water, rather than the common practice of grabbing them by the tail, mouth or under the gills. Live fish should not be handled simply to photograph them. Fish could be killed rather than returned to the water, debilitated and injured, and hook removal could be postponed until after killing. However, the fish will still suffer considerable pain and distress, even with these refinements.

Shooting Huge numbers of game birds are reared and released for the sport of shooting, and it is likely that many will be injured rather than killed outright, especially where shooters have less training and experience.

Infectious Diseases Introducing infectious diseases can be devastating. An example is the outbreaks of myxomatosis in the UK and Australia during the 1950s, believed to have killed millions of rabbits. Many will have suffered disease symptoms for at least several days before finally dying. The virus was deliberately introduced to control rabbit numbers.

Free-living, incidentally harmed wild animals

Activities that support human lifestyles often injure animals.

Transportation Systems Transportation systems injure and kill many animals. Collisions with vehicles are estimated to kill 10–100 million birds and mammals annually on Europe's roads. The large accidental oil spills that injure thousands of birds are widely publicised. However, the many smaller, routine oil emissions that are low-profile harm an even greater number of wild animals. Fragmentation of habitats prevents animal populations from migrating normally, meaning they cannot access the resources they need to remain fit at different times of day and seasonally.

Agricultural Practices Agricultural practices such as ploughing injure and kill many small mammals, especially rodents such as voles and mice. They are injured by equipment blades, suffocated in their burrows under compacted soil, or preyed upon once the covering vegetation is removed. Overgrazing or repeated cultivation harms the habitat quality for animals. Widespread use of chemicals, such as pesticides and herbicides, can directly harm animals in the areas where they are applied and also harm those further afield due to the pollution they create when they leach into waterways. An example is high morbidity and mortality in fish due to the creation of hypoxic zones.

Waste Disposal Animals regularly become entrapped in discarded waste items, such as plastic bags, bottles and nets or are harmed by swallowing them.

Non-native Predator Species Introducing non-native predator species significantly disrupts ecological systems. One of the most ubiquitous is the domestic cat, which is estimated to injure and kill approximately 100 million birds and mammals annually in Great Britain. Wild animals are commonly admitted to wildlife hospitals because of cat attacks.[169]

Captive wild animals

Limited species-specific knowledge exists, designing appropriate enclosures is challenging and animal populations must be actively managed.

Limited Species-Specific Knowledge The sheer diversity of species and limited knowledge about specific species make preventing and identifying problems challenging. For example, signs of pain are ill-defined for many species, and detection is made even more difficult because many animals are prey species and, therefore, show few overt signs.

To mitigate, veterinarians and keepers can take a proactive approach, working closely together to identify and investigate subtle behaviour changes. Regular visual examinations, physical examinations under anaesthesia, and routine disease and parasite screening aid in the early detection of problems. The stresses of such interventions, particularly in non-domesticated species, need to be weighed against the overall welfare benefits. Refinements include ensuring the enclosure has an area where animals can be separated, handled and examined humanely. Enclosure design should also prevent animals from escaping, protecting them, other animals and the general public from injury. Controlling access to the enclosure is also important for biosecurity.

Other important welfare issues include population management, fitness for release, geriatric animals and effective triage.

Population Management Breeding needs to be managed to avoid pain, injury and disease. Uncontrolled breeding can lead to overpopulation, overcrowding and the associated harms of disease transmission, stress and injury from social tensions, fights among group members, and behavioural restrictions. Different population management strategies are used – either 'breed and cull' or contraception – with different ethical and welfare trade-offs (see F5).

Zoos actively promote the breeding of underrepresented species. Breeding needs to be planned and monitored to maintain genetic diversity and avoid health and welfare problems associated with inbreeding. Invasive reproductive procedures such as artificial insemination and embryo transfer may be used, along with associated pain and distress, to combat low fertility rates. However, the root cause of failure to breed naturally is often a suboptimal environment, which should be addressed instead.

Fitness for Release Limited species-specific knowledge also applies to rescue-rehabilitation centres.

It is crucial that fitness for release back into the wild is properly evaluated. Success should be judged according to post-release welfare and survival, not just the number of animals released. Such monitoring rarely occurs, and few studies are available to inform best practices.

Post-release welfare can be optimised by careful planning. Factors such as release site location and time of year can make a difference, for example, determining whether sufficient food will be available and avoiding release during the breeding seasons for species that become more territorial at these times. It is also important to consider the impact of the release on other free-living wild animals, such as whether new arrivals will disrupt social groupings and whether new diseases will be introduced, especially by translocated animals. Some animals may require a 'soft' release, with initial resources provided by humans, as they adjust, rather than a 'hard' release where they are set free without support. Veterinary professionals should be involved at this stage, not just the initial admission triage.

Geriatric Animals Animals are living longer in captivity due to improved healthcare, which can be a positive welfare indicator; however, this also presents specific challenges. Geriatric diseases such as osteoarthritis, dental, heart and renal disease and cognitive decline are on the rise. Often, captive wild animal facilities are motivated to cull animals for population management and collection planning purposes, but are reluctant to euthanise chronically sick or suffering animals, especially if they are a charismatic species or popular with the public or staff. To mitigate this, veterinary professionals can provide important guidance, and facilities should develop ethical committees that include external experts to ensure balanced decision making.

Triage Effective triage that prioritises welfare is important when wild animals are presented to wildlife hospitals, other rescue rehabilitation facilities, and veterinary practices. This should involve veterinary professionals. Often, there is public pressure to admit and treat injured wildlife, and doing so can create good public relations for the practice or centre. However, this should be resisted where inappropriate. Instead, there is often an opportunity to educate citizens on welfare, giving good reasons why euthanasia is often in the animal's best interests.

F4. Freedom from fear and distress

Pain, injury and disease are generally associated with unpleasant feelings such as fear and distress, but they can also occur independently.

All free-living wild animals

All free-living wild animals will suffer degrees of fear and distress when trapped, either intentionally or incidentally. This fear is associated with being held in one place, unable to move, and feeling exposed to predation.

Captive wild animals

Lack of Domestication Wild species have not undergone domestication and are genetically less adapted to close contact with humans than domesticated species. As a result, they experience heightened stress and fear in captivity. In rescue-rehabilitation facilities, fear of humans during the treatment and rehabilitation phase can cause distress and immunosuppression, slowing healing and predisposing to further disease.

Exposure to Humans Zoos require visitors to be able to view animals. There is a tension between the animal's best interests and those required of the public. Being on view is not necessarily a problem for the animal, but knowing or perceiving they are on view often is. Refinements are needed to mitigate. Studies suggest that large cat species, such as jaguars, feel stressed when viewed by people positioned above them. Therefore, cat viewing areas should be at ground level or below the cats. Excessive viewing of primates creates anxiety, often manifested as abnormal behaviours such as hair plucking. This can be mitigated by providing animals with vegetation and furniture in the enclosure, making them feel sheltered and out of sight.

Habituation to People Wild animals permanently kept in zoos and sanctuaries probably benefit from habituating to people, but this could be detrimental for animals that are to be released into the wild. Rescue-rehabilitation facilities need to determine the right level of contact – too much habituation to humans, and the animals will not be wary enough to cope well once released.

Presence of Other Species Animals usually have well-developed senses of hearing and smell, and can sense vibrations and movements. They will likely be aware of other species housed on the same site, despite being physically separated from them, and this may cause them fear and distress.

Standards of Care The person caring for the animals is a key determinant of their welfare. Standards of care are likely to vary greatly. Across rescue-rehabilitation facilities, carers vary from enthusiastic laypeople with little training and limited resources to highly trained personnel with considerable expertise working in well-resourced facilities. In the former, intent may be good, but not always balanced with knowledge and skills.

F5. Freedom to express normal behaviour

Human interactions and actions impact the normal behaviours of many wild animals in a way that often harms their welfare. Knowledge is limited compared with that of domesticated species.

Free-living, deliberately targeted wild animals

Trapped adult animals that are separated from their young cannot express parental behaviours that they are highly motivated to perform. In addition, their young will suffer because they do not have access to parental care-giving behaviours. Non-target animals accidentally trapped will also be affected.

Free-living, incidentally harmed wild animals

Human activities unintentionally disrupt a range of behaviours important for the welfare of both terrestrial and aquatic animals. Examples include transportation systems that fragment habitats and disrupt migratory behaviours that animals are often strongly motivated to perform. Artificial lighting at night can confuse birds, disrupting their normal migrations, and can disrupt the circadian rhythms of animals generally, impacting their normal behaviour. Noises and vibrations can disrupt communication among animals.

Captive wild animals

Key welfare issues include a lack of domestication, enclosure design, environmental enrichment, abnormal behaviours, population management and hand-rearing of the young.

Lack of Domestication Wild animals in captivity face unique challenges. Undomesticated animals, even those born and raised in captivity, are less adapted to captive environments and close contact with people than domesticated species are. A degree of adaptation can occur when enclosure designs and management practices are carefully crafted to meet the animals' needs, and they learn new behaviours and skills through reward-based training.

Enclosure Design Enclosures need to provide for important species-specific behaviours such as flying, climbing, swimming or foraging, not just basic comfort behaviours. Barren, monotonous environments cause boredom and frustration, which can progress to depression and apathy, otherwise known as 'learned helplessness'.

Enrichment Provision is only enriching if it is meaningful to the animals, for instance by encouraging the diversity and time budgets of natural behaviours in the wild. The level of challenge offered is important. Too little change and stimulus will bore them, and too much can create stress and fear. Allowing animals choices and active engagement in problem-solving benefits them. An example is the use of underwater enrichments, such as puzzle feeders and hoses, to reduce the barrenness of traditional underwater glass and concrete environments for cetaceans. This encourages problem-solving and normal predatory behaviours, and is superior to traditional enrichments, such as floating items like balls, which lead them to spend significantly more time at the water surface than their wild counterparts.

Modifying routine husbandry practices, such as feeding, can make enrichments more practical. For example, feed could be scattered unpredictably by varying the time, frequency and location.

Sufficient enrichment must be provided for all the animals in the group to prevent competition and aggression. Enrichments also must be safe, with a low risk of injury, choking, toxicity or entanglement. Importantly, their ongoing usage needs to be monitored. Preferences will vary among individuals, and according to ages and stages of development.

Many wild animals can be trained using positive rewards. An animal can be trained to return to a designated holding area within its enclosure at specific times of the day or in response to specific signals. Such training is useful for multiple reasons, including allowing keepers to set up more complex enrichments, such as burying, hanging or hiding objects under logs or piles, without risking their own safety.

Environmental enrichment will benefit species of all taxa. However, implementation can be difficult due to insufficient knowledge and resources, which is unsurprising given the vast diversity of species. Recreating the size and complexity of many natural environments where free-living animals thrive is impossible, so decisions must be made about which aspects matter most to the animal. For example, some species have extensive home ranges and territories and migrate over many miles, exploring and using navigational skills. The bottlenose dolphin can travel tens of kilometres daily, with home ranges exceeding 100 square kilometres.[170] The terrain occupied by the Arctic fox may appear very barren to humans, but it will provide enormous stimuli for the fox that people cannot perceive. Many captive environments are likely to remain suboptimal, despite the best enrichment attempts.

Abnormal Behaviours Abnormal behaviours indicate that animals are experiencing problems. They are observed in many captive species, including bears, canids, elephants, primates, ruminants, birds and even fish and reptiles. Examples include stereotypies such as pacing and head tossing, and behaviours where animals self-harm or harm others in the group, such as feather plucking and overgrooming. Simply interrupting or preventing abnormal behaviour can worsen welfare, as the behaviour itself may play a role in alleviating anxiety. Instead, the root cause – usually some deficiency in the animal's environment – should be identified and addressed.

Population Management Different approaches are used to manage populations. In 'breed and cull' strategies, offspring are maintained until the age of natural dispersal, at which point they are humanely culled. This has the advantage of allowing normal reproductive behaviours to be expressed, which generally improves social cohesion among animals in a group. It also allows for the expression of maternal behaviour, which may benefit the mother. An alternative approach uses contraception to suppress natural breeding behaviours. Here, culling is avoided. However, there is potential for chronic behavioural frustration and reproductive pathology.

For some species, breeding is desired because of low population numbers. The fact that some species fail to breed is often a sign that their needs are not being met.

Hand-rearing Removing young animals from their mothers and hand-rearing them often occurs when the mother fails to express normal maternal behaviour or the animal has a high conservation value. However, there is now considerable scientific evidence that a lack of maternal contact is a major postnatal stressor that damages the development of the young. This stressor can cause abnormal brain development and psychological disorders. Again, the root cause of poor maternal behaviour, usually an environmental deficiency, should be addressed. Where natural breeding occurs, there is more likely to be normal maternal behaviour.

Hand-reared young in rescue-rehabilitation facilities are unlikely to develop the physical and mental skills to cope well in the wild. Only some behaviours are innate, with many learned from parents. Many behaviours will also require practice if they are to develop sufficiently and if the animal is to have the physical conditioning to perform them consistently. This is an issue for animals released into the wild from rescue-rehabilitation centres – they have less ability to cope than their counterparts born and raised in the wild.

Welfare Assessment Protocols

Protocols are being developed for some species of wild animals in zoos and aquaria. Examples include the development of a welfare assessment index for captive bottlenose dolphins.[171] The Association of Zoos and Aquariums (AZA) Nutrition Advisory Group is compiling body condition scoring systems developed for various captive wild species.[172] However, much more research is needed on score validity and to expand the range of species covered.

Advocacy examples

Free-living, deliberately targeted wild animals

- Position statements and guidance produced jointly by the British Veterinary Association (BVA) and the British Veterinary Zoological Society (BVZS) in the UK cover the control of free-ranging wildlife (2021), the use and sale of snares for the control of free-ranging wildlife (2022) and the use and sale of rodent glue traps (2021). They are shaped by the international consensus principles for ethical wildlife control developed by Dubois and others[173] that the BVA adopted after learning about them at a wildlife control workshop in 2015.
- In the UK, the Wild Animal Welfare Committee[174] provides independent advice, evidence and educational resources about the welfare of free-living wild animals.
- The AnimalKind Accreditation Program, launched by the British Columbia Society for the Prevention of Cruelty to Animals (BCSPCA), is the first welfare accreditation scheme for wildlife and rodent control companies. It is based on the international consensus ethical principles of Dubois, but with simplified language for easier understanding.[175]
- Veterinary professionals can promote better animal protection through regulation. They can reflect on their views of wild animals commonly labelled as 'pests' and educate the public to try to garner their support. Research is also essential to ensure actions are grounded in scientific evidence.

Free-living, incidentally harmed wild animals

- Many environmental sustainability initiatives can support wildlife welfare and the well-being of humans and animals, in line with the 'One Health' philosophy, which recognises the co-dependency between the health of animals, people, plants and the environment.
- In the UK, the BVA initiative, Green Team Vet,[176] supports the veterinary profession to be more environmentally friendly. They collaborate with Vet Sustain, which was founded in 2019 by a group of volunteers, and provides useful educational resources on a wide range of topics, from 'greening' veterinary practices and advising on regenerative farming to navigating conversations with clients about sustainable pet ownership.

- Food Animal Initiative (FAI)[29] is a commercial enterprise that supports farmers in implementing farming systems that not only provide significant animal welfare benefits for the animals farmed but also for the environment, which in turn supports the welfare of free-living wild animals and human health.
- Veterinary professionals can support other initiatives likely to improve wildlife welfare and research their effectiveness. Examples include the development of wildlife crossings, which are human-built infrastructures, such as bridges or tunnels, designed to help wildlife safely cross roads and reconnect habitats that have been fragmented by infrastructure. In Europe, the Saferoad project is studying these.[177]
- Cat owners can be educated on how to limit the predatory activities of their cats whilst also maintaining cat welfare, using advice from organisations such as International Cat Care.

Captive wild animals

- The British Veterinary Zoological Society (BVZS) has produced good practice guidelines for wildlife rehabilitation centres.[178]
- The British Wildlife Rehabilitation Council (BWRC) is a UK-based charity seeking to improve the welfare of native wildlife casualties by supporting and educating workers in the field.
- At the international level, the Global Federation of Animal Sanctuaries (GFAS) supports the development of accreditation systems for sanctuaries, rescue-rehabilitation centres and wildlife hospitals and provides guidance on raising standards.
- The Association of Zoo & Exotic Veterinary Nurses (AZEVN),[179] formed in 2016, is run by veterinary nurses and provides continuing education focused on the zoo and exotic sector.
- The Wildlife Disease Association[180] is an international scientific society of wildlife professionals, including veterinary professionals, involved in wildlife health and conservation from over 70 countries. Veterinary professionals can bring an understanding of animal welfare and its relationship with health and conservation to discussions.
- Various national and international organisations representing zoos have increasingly taken animal welfare concepts on board, including them as part of their guidelines and standards against which facilities are accredited. For instance, since 2018, the revised Secretary of State's Standards of Modern Zoo Practice,[181] which provides minimum standards for the licensing of UK zoos, confers a duty of care on all animal caregivers based around the Five Welfare Needs framework. The European Association of Zoos and Aquaria (EAZA) has an explanatory webpage on its approach to improving welfare.[182] The World Association of Zoos and Aquariums (WAZA) has established clear targets and timelines for members to implement an animal welfare evaluation process. Veterinary professionals can help organisations understand how the welfare concepts apply in these contexts.
- An important area of advocacy outside the scope of this chapter is where non-domesticated animals are kept as pets by the general public.

Laboratory Animals

16

Abstract

The core welfare concepts discussed in Part 1 are applied to laboratory animals. Some common welfare hazards are listed. Common welfare outcomes are identified and structured according to the Five Freedoms framework. Examples are provided of how veterinary professionals have advocated for laboratory animal welfare or could be more involved.

Introduction

Laboratory animals are generally defined as animals used or kept for scientific experimental procedures. They are a high-profile moral concern in many countries. Citizens are generally aware that animals are harmed to provide information that will benefit other animals and humans in the future. This contrasts with veterinary practice, where shorter-term harms, such as surgery, are often (though not always) performed for the longer-term benefit of the same individual.

In this chapter, a scientific experimental procedure is defined as a procedure done on an animal as part of a scientific study that causes the animal to experience pain, suffering, distress or lasting harm equivalent to or higher than the introduction of a needle in line with good veterinary practice (this mirrors the UK legal definition of a 'regulated procedure' under Animals (Scientific Procedures) Act (ASPA) 1986). This, therefore, covers a very wide range of procedures. Emotive campaigns designed to stop the use of animals in research often fail to capture the nuance – the numbers and species of animals, types of research and procedures, and the development of safeguards – something this chapter seeks to highlight.

Scale and type of animal usage

Laboratory animal usage varies globally, and accurate data are difficult to obtain for many regions. However, the UK and EU countries have some of the most

DOI: 10.1079/9781786391582.0016

stringent regulations, and their official bodies publish reports that provide insights into the scale and type of their animal usage. Figures vary annually and between regions, but some general 'ballpark' figures are as follows. Millions of animals are used in experimental scientific procedures each year (in recent years, between 2.7–3.4 million per year in the UK,[183] and around 9–10 million per year across the 27 EU countries). In these regions, mice are the most commonly used species (around 50–70% of animals), followed by fish (around 15–30%), rats (around 6–12%) and then birds (around 4–9%). In addition, many more animals are used in 'breeding and creation procedures', creating genetically modified animals for future use in scientific studies. The majority of these are mice.[184]

Mice are commonly used because they are easy to maintain due to their small size and high reproductive rate. They are also susceptible to many genetic diseases that afflict humans and are the main species genetically altered to produce specific research models. In the UK and EU, most laboratory animals are purpose-bred.

Research types can be broadly categorised as follows:[184]

- **Basic research**: expands knowledge of living organisms and the environment without focusing on solving a specific problem.
- **Applied research:** focuses on specific problems, such as investigating the causes and prevention of diseases and developing treatments.
- **Regulatory research**: largely tests the safety of compounds such as pharmaceuticals or agrochemicals for humans, animals and the environment so that appropriate warnings and information can be issued alongside their usage.
- **Other**: examples include specific surgical training and education.

In the UK, the severity of procedures is banded as follows:[184]

- **Mild**: pain and suffering are slight or transitory.
- **Moderate**: significant but non-life-threatening disturbances to an animal's normal state, for example surgeries performed under anaesthesia with adequate post-operative pain relief.
- **Severe**: significant departure from the animal's normal health and well-being. This includes long-term diseases where the animal requires assistance to eat and drink or has major behavioural dysfunction.
- **Non-recovery**: the entire procedure is performed under general anaesthesia, and the animal does not regain consciousness.

In Great Britain, in recent years, around half of all experimental procedures were basic research, around one-quarter were applied, around one-fifth were regulatory, and around 2% were for other purposes. Under 5% of all experimental procedures were classified as severe, approximately one-quarter as moderate, around half as mild, and less than 5% as non-recovery. The Home Office updates figures annually.[184]

Severity levels cover the peak harms or the cumulative severity of the scientific procedure, but not harms associated with other events in the animal's life outside of the scientific procedure, such as breeding, housing, husbandry, transport and killing methods. This also deserves attention, as only a fraction of the animal's lifespan is spent undergoing a scientific procedure, and many animals are never used at all.

Laboratory animal usage and societal morals

The current use of animals in experiments is heavily influenced by societal morals and assumptions, which are further reinforced by laws. These societal norms and legal frameworks serve as the 'backdrop' against which further ethical discussions and developments will occur, and they are likely to vary significantly across different countries and cultures.

The current UK approach, much of which the EU shares, is largely utilitarian. When deciding whether laboratory animal usage is justified, a harm–benefit analysis is undertaken, with the likely benefits to humans and other animals weighed against the harms to the laboratory animals. However, some procedures are considered so harmful that they should not be done at all, regardless of the benefits to others. Any harm should be minimised, and animal use is illegal if a non-animal alternative is available. Some species receive a special dispensation because our culture favours them; dogs, cats, horses and non-human primates must not be used unless there is no alternative. In the UK, fewer than 1% of experimental procedures involved dogs, cats, non-human primates, or horses, and none were used in breeding and creation procedures. The great apes, such as gorillas, chimpanzees and orangutans, are considered so similar to humans that their usage is banned. Some reasons are considered too trivial to warrant animal usage. For example, in the UK, no animals are used to test cosmetics or finished household cleaners.

The EU and the UK have strict regulations governing the use of laboratory animals. However, laboratory animals are exempt from general animal welfare laws, and many procedures they undergo would be offences if performed in a non-experimental setting. In the UK, several types of regulatory tools are made good use of, mandated by law (Animals (Scientific Procedures) Act 1986):

- Species protected by law are clearly defined (non-human vertebrates and cephalopods, including certain immature forms).
- All projects are reviewed by an Animal Welfare and Ethical Review Body (AWERB) to decide whether they can be ethically justified. The composition of the Committee will determine the quality of the review.
- Extensive use is made of licensing that includes detailed requirements. Project licences cover each programme of work, personal licences cover each researcher who must undergo training and demonstrate competency, and the research establishment must also be licensed.

- At the research establishment, a veterinary surgeon (Named Veterinary Surgeon) must advise on the health, welfare and ethics of experimental animal usage during all project stages and on the welfare of animals outside of scientific procedures.
- The Home Office ensures the law is implemented and enforced through routine inspections by Home Office Inspectors who also investigate reported infringements.
- The Home Office issues a detailed Code of Practice for standards of care and accommodation.
- Certain data must be transparent; reports on numbers and types of scientific procedures and their severity must be published and laid before Parliament.

Husbandry

The requirements of good science determine which animals are bred and created, and how they are kept. Researchers aim to minimise variation within and amongst studies, increasing the reproducibility and comparability of results. This requires standardised animals kept under highly controlled conditions. Often, animals must possess specific genetic or microbiological characteristics or be disease-free.

Genetically altered animals

Genetic alterations are made in laboratory mice to meet scientific needs, going beyond conventional farming and companion animal breeding practices. Many are done to produce models for specific human diseases. Here, specific genes may be eliminated, replaced with an altered form or overexpressed. Suitable mice are then bred, passing changes to the next generation.

Creating genetically altered mice impacts the welfare of large numbers of animals, with 50 to several hundred mice used to produce a single strain. Also, some genetic alterations are deliberately harmful, such as those linked to neurodegenerative diseases. Other gene alterations unintentionally create harm. For example, mice in which a growth factor receptor gene was removed exhibited severe abnormalities, including skeletal defects and profound deafness, that were not anticipated.[185] Some genetically altered mice may appear grossly normal, but might have subtle defects detrimental to their welfare.

Relationship between welfare and sound science

Just as the welfare and production of farmed animals can improve in tandem up to a point, so can the welfare of laboratory animals and the quality of science. Russel and Burch, two scientists who worked alongside the Universities Federation for Animal Welfare (UFAW) to review laboratory techniques and their ethical implications, emphasised this in their book titled 'The Principles of Humane Experimental Technique', published in 1959.[55] Factors that reduce welfare can also make scientific results less reliable. For example, if animals are kept at uncomfortable temperatures, this can alter their metabolic rate and skew the results of drug toxicity studies. Poor handling during procedures can increase anxiety levels and stress responses in animals, impacting data. Russel and Burch explained how science could be more humane without thwarting medical and scientific aims. They captured this concept as the 3Rs – Replace, Reduce, Refine (see Chapter 4). This has helped people move away from the polarised positions of wanting to stop animal use completely or wanting to keep animal use unchanged. The 3Rs initially had little uptake but have gradually gained popularity. Today, they are widely recognised internationally as a framework for improving laboratory animal welfare and are mandated by law in some countries. For instance, a national centre for replacement, reduction and refinement of animals in research (NC3Rs)[186] was established in the UK in 2004. Public support for animal research appears conditional on proper implementation of the 3Rs.

It is important to note that the 3Rs are a crucial aspect of ethical review, but do not encompass the entirety of ethical analysis. They serve as a 'utilitarian' framework aimed at minimising harm once it has been determined that an experiment can proceed. However, engaging in a thorough ethical discussion about whether the study can be justified in the first place is important.

The 3Rs Framework

- **Replacement**: Using non-animal alternatives or animals that are less sentient.
- **Reduction**: Fewer animals are used due to good experimental design, appropriate statistics and the use of the correct animals and strains.
- **Refinement**: Modifying procedures so that they are less harmful, and improving housing and husbandry to reduce any associated suffering.

Replacement and reduction mean fewer animals suffer overall. Refinement concerns the severity and duration of the suffering of the animals used, that is, their welfare.

Some replacements are easier to develop than others. Many chemical toxicity tests are standardised, making developing, validating and implementing alternatives easier. Examples include *in vitro* methods that use cells or animal tissues instead of live animals, specialised computer modelling and data sharing. Other replacements are more challenging. For instance, relatively large numbers of live animals are still needed to test biological products such as vaccines, as their exact composition and properties can be challenging to control.

Refinements involve altering techniques to minimise suffering and optimise welfare, aiming to meet the system's 'welfare potential'. A holistic welfare assessment framework, such as the Five Freedoms, ensures that all areas important for welfare are considered. It must be borne in mind that seemingly 'normal' animals that have been altered by genetic modification might have some altered needs.

Common welfare hazards

For laboratory animals, there is huge variation in factors such as housing and husbandry within and outside the scientific procedure, the type and severity of the procedure, animal factors (species, strain and age), and human factors (the skills, attitudes and resources of the human carers). Therefore, to narrow the context, examples are given from laboratory mice.

- Feeding regime.
- Environmental conditions (temperature, lighting, noise, odours).
- Level of enrichment.
- Social grouping.
- Level of pain relief.
- Handling quality.

Welfare hazards are the resources (inputs) that can lead to poor welfare outcomes. The likelihood of poor welfare outcomes and their severity (i.e. the risks) can vary. Severity encompasses both duration and intensity. Common welfare outcomes are identified below, structured by the Five Freedoms framework.

Mouse welfare is determined by how well mice can adapt, both physically and mentally, to their environment (see p.8). Relevant descriptions of 'free-living' mice are provided throughout to highlight the adjustments they must make while in human care. Here, 'free-living' is defined as mice living freely in an environment with the resources for which they have evolved and making their own choices. However, such comparisons provide only part of the welfare picture; welfare science (see Chapter 2) is essential for a proper understanding of the welfare impact of different husbandry systems.

Key determinants of welfare are the knowledge, skill and motivations of the people caring for the mice, as well as whether they have sufficient resources (see pp. 24–25). Where these factors are optimal, welfare will be as good as it can be within a given husbandry system. However, there are often persistent welfare issues that are challenging to resolve without altering the husbandry system itself.

Common welfare outcomes

F1. Freedom from hunger and thirst

'Free-living' mice eat a wide variety of foods, including seeds, grains, roots, leaves and fruits. They have large incisor teeth that grow continuously and are worn down by gnawing on hard materials and grinding food. In contrast, laboratory mice are typically fed pelleted food that meets their basic nutritional needs but doesn't require much chewing, which can lead to overgrown teeth and dental issues. Pain caused by experimental procedures can also make it difficult for them to eat and drink.

Refinements

Refinements to mitigate hunger and thirst include providing appropriate analgesics to alleviate pain, and offering gnawing materials to promote dental health and satisfy natural behaviour. (Note that these are not necessarily done routinely – scientists will identify any impediments that refinements pose to experimental procedures and data collection, which veterinary professionals must then navigate when advocating for the animals.)

F2. Freedom from discomfort

'Free-living' mice occupy various habitats and adapt their behaviour based on environmental conditions. They are sensitive to temperature changes due to their large surface-area-to-volume ratio. They require significant energy to stay warm in cold temperatures and burrow to stay cool in hot temperatures. They are active during twilight or at night and are highly sensitive to light. They can hear high-frequency ultrasound, which is inaudible to humans. Laboratory mice have limited choices and can easily become too hot or too cold. The immediate environmental temperature is influenced by factors such as the cage material (plastic or metal), bedding and the number of mice in the cage (Figure 16-1). Altered light duration and intensity can disrupt circadian rhythms, affecting physiology and behaviour. Excessive background noise may also cause discomfort.

Refinements

Refinements to mitigate discomfort include the following:

- Research institutions should monitor and address the internal cage environment at mouse level because conditions here can differ considerably from those of the wider room containing the cages, which are more commonly monitored.

Figure 16-1. Typical container used to house laboratory mice.

- There are various ways to reduce light intensity at cage level, such as dimming lights when personnel are not working, providing a shelf over the top row of cages to block light, and offering mice tunnels for resting and sleeping.
- Background noise can be minimised to prevent discomfort and communication disruption among mice. It is important to be aware that many noises inaudible to humans can disrupt mice, such as noise produced by white goods.

F3. Freedom from pain, injury and disease

Laboratory mice undergo various experimental procedures, including surgery and other invasive techniques, that cause pain and injury. They present unique challenges for anaesthesia due to their small size and limited access to blood vessels. Research staff may have varying expertise and may work alone, leading to suboptimal anaesthesia, potentially resulting in hypothermia, hypoxia, hypercapnia or overdose.

Refinements

Refinements to mitigate pain, injury and disease include good communication with and input from the veterinary professionals in the research establishment, ensuring staff are trained in welfare assessment and have the authority to act on findings. Suffering can be reduced by adopting 'humane endpoints', that is, euthanising mice when they show signs indicating they will later die rather than waiting for them to actually die.

F4. Freedom from fear and distress

Mice are prey species. They use scents in the air and on physical objects to detect predators. Laboratory mice are often handled by humans whom they perceive as large predators. Due to their small size, mice are easily overpowered; for instance, they are commonly lifted by the tail, causing them pain and anxiety.

Refinements

Refinements to mitigate fear and distress include handling mice gently to habituate them. Instead of being lifted by the tail, they can be lifted in a tunnel they have entered or using cupped hands. Rats should be housed in a separate air space as they are a natural predator of mice.

F5. Freedom to express normal behaviour

'Free-living' mice exhibit various behaviours such as exploration, foraging, burrowing, grooming and nesting. They are skilled at jumping, running and climbing. They are social animals that live in stable groups, often consisting of one male, several females and their young. Social hierarchies form, which limits conflict and aggression. Mice heavily rely on scent and pheromones to communicate and maintain stability within their colony. Much of their behaviour revolves around marking and defending territory from intruders. In contrast, laboratory mice are often kept in small plastic 'shoebox style' accommodations with limited resources, thwarting their natural behaviour and leading to boredom and frustration. Altered social groupings may trigger aggression. Removing scent marks and pheromones during cage cleaning also triggers insecurity, aggression and fighting as mice reestablish their territories. Subordinate mice may experience 'barbering', that is, dominant mice chewing their whiskers.

Refinements

Refinements allowing normal behavioural expression include environmental enrichment. This should be considered essential for normal behaviour, physiology and immune function, not an optional luxury. Their space should contain solid flooring, nesting material, bedding, objects for climbing and refuges for retreat, with enough resources for all group members. Group mates should be carefully chosen. For instance, male mice can be aggressive unless raised together from birth. When cleaning the cage, leaving a small amount of soiled bedding behind helps preserve the scent environment, so mice do not need to completely re-establish social hierarchies when they re-enter.

Inherent welfare problems

Scientific needs often limit refinements. For example, experimental procedures may require restricting food and fluids, causing hunger and thirst. Changing anaesthesia and pain relief mid-study can benefit welfare but introduce unwanted scientific variation. Opaque walls in housing can make mice feel secure but impede scientific observations.

Welfare assessment protocols

Reliable scoring systems are needed to assess welfare during and outside of scientific procedures. They should capture welfare over the animal's lifespan, including when sourced, transported, housed in the research establishment, used in scientific procedures and killed. Desirable general features of scoring systems are set out in Chapter 3, Practical Welfare Assessment. However, developing protocols for use during scientific procedures is more complicated than in many other contexts. Procedures vary greatly, have very different impacts, and there are more unknowns. For example, signs of distress may vary according to the particular organ affected. Often, pilot studies are necessary, and protocols may need to be modified as the experiment progresses. Where animals have been genetically altered, this further increases uncertainty, as their needs may differ from those of non-modified animals. In addition, the large number of strains created makes it difficult to evaluate them all properly. The NC3Rs working party recommended developing structured welfare assessments to compare the welfare of genetically altered mice with that of non-altered mice.

Advocacy examples

- Veterinary professionals can make valuable contributions to the welfare of laboratory animals, whether they work directly in this field or not, provided they become well-informed.
- Veterinary professionals working in scientific establishments can directly contribute valuable animal welfare insights to ethical review processes, experimental designs and ongoing monitoring of animals. They can also advise on the welfare of the many animals bred and kept outside of experiments.
- 'Understanding Animal Research'[187] is a reliable information source for the UK. This organisation launched the 'Concordat on Openness on

Animal Research' in 2014, committing signatories to enhance their communication about animal use in research. The Concordat has been signed by numerous organisations, including the Laboratory Animal Science Association (LASA), and has inspired the creation of similar transparency agreements in other countries.

- Resources focusing specifically on welfare include the NC3Rs[186] and UFAW webpages, as well as the 'UFAW Handbook of the Care and Management of Laboratory and Other Research Animals'.
- National and international veterinary professional associations that provide useful information include the Laboratory Animal Veterinary Association (LAVA) based in the UK, the European Society of Laboratory Animal Veterinarians (ESLAV), the European College of Laboratory Animal Medicine (ECLAM) and the American College of Laboratory Animal Medicine (ACLAM).
- Position statements by veterinary associations to support and improve animal welfare include the ACLAM position statement on adequate veterinary care,[188] and guidelines for veterinary care produced jointly by FELASA, ECLAM and ESLAV.[189]
- Some organisations provide students with the opportunity to learn about animal research and the various veterinary roles associated with it. An example is the extramural studies programme run by LAVA in the UK.[190]
- The EU and the UK[191] have some of the most stringent laws globally controlling the usage of laboratory animals, setting minimum standards for the housing and care of animals, training of personnel, and encouraging the development of alternatives to animal usage. Standards are evidence-based, informed by EFSA reports.[16] There is ongoing review.
- Regarding implementing the 3Rs, the EU has established a reference laboratory for alternatives to animal testing that supports research and development in this area and shares knowledge.[192]
- The Association for Assessment and Accreditation of Laboratory Animal Care (AAALAC) is a charity that works internationally to promote the humane treatment of animals in science through voluntary accreditation and assessment programmes, with many companies, universities, hospitals, government agencies and other research institutions taking part.[193]

Humane Killing

17

Abstract
The core welfare concepts from Part 1 are applied to situations where animals are intentionally killed. Common welfare hazards are listed. Common welfare outcomes are identified and structured according to the Five Freedoms framework. Practical welfare assessment protocols are referenced. Examples are provided of how veterinary professionals have advocated for welfare at the time of killing or could be more involved.

Introduction

Animals are killed for various reasons, including alleviating suffering, producing food, for sport, controlling populations, managing infectious and zoonotic diseases, and removing animals that are performing poorly.

The killing process inevitably involves inflicting extreme physical or physiological injury or insult to the animal's body, which would generally cause considerable suffering if consciously experienced. Therefore, from a welfare perspective, the ideal killing method causes the animal to become immediately unconscious. If consciousness is lost more gradually, then this should be without pain, distress or suffering. The animal should then remain unconscious and insensible until they are dead.

Killing animals usually involves handling and transporting them, so welfare during these operations must also be considered in assessing overall welfare associated with killing. The killing process also needs to consider productivity and human welfare. In the case of depopulation for disease control, the speed of operation and biosecurity are crucial to prevent the further spread of disease. People involved need to stay safe, and the process needs to be efficient.

To optimise animal welfare during handling, transport and killing, it is essential for people to understand how animals perceive their environment and why they react in the way that they do.[194] This includes understanding the importance of social peers, what motivates animals to move, how they learn,

DOI: 10.1079/9781786391582.0017

and what they have learned from prior experiences. It is essential to translate this knowledge into practical action. Those involved in planning and monitoring should be skilled in interpreting animal behaviour as an indicator of their emotions. Although resources like well-designed handling facilities can be beneficial, the most significant factors affecting animal welfare are often the skills, knowledge and attitudes of the individuals involved.

This chapter focuses on the intentional killing of commonly farmed species in the UK and EU countries.

Operational systems

Operational details are first described to provide sufficient information for engaging with the subsequent welfare analysis.

Humane killing

In this chapter, killing methods are considered humane if they cause loss of consciousness and sensibility without pain and maintain this state until the animal is dead. This definition aligns with UK and EU[196] laws that protect farm animals at the time of killing. Related operations, such as handling, transport and husbandry in the lairage, must also respect the animals' needs. While striving for these ideals, few killing methods are perfect and 'humane' can be a relative term. The killing method is often judged according to whether it causes 'avoidable' or 'unnecessary' pain or suffering. Interpretations vary according to species and context, determined by socioeconomics, logistics and morals.

This chapter uses the term humane killing rather than other terms associated with killing, such as slaughter or euthanasia, to avoid confusion. To illustrate, under EU law,[196] slaughter means the killing of animals intended for human consumption, whereas the World Organisation for Animal Health (WOAH) defines it as any procedure that causes the death of an animal by bleeding. For some people, slaughter has negative connotations, such as inhumane, immoral and mass killings. The word euthanasia derives from the Greek language for a good death, that is, 'eu' (good) and 'thanatos' (death). However, some authors argue that the term should be morally loaded and used only for the deaths of animals that are killed humanely 'for their own good' because, for example, they are suffering in a way that cannot be otherwise alleviated, rather than for the humane killing of healthy animals that are unwanted or are to be used for food.

Stunning

The term 'stunning' is useful when discussing humane killing as it captures a core principle essential for welfare – it is defined as any intentionally induced process which causes loss of consciousness and sensibility without pain, including any process resulting in instantaneous death.[196] The terms 'consciousness' and 'sensibility to pain' are used interchangeably in this chapter as their meanings overlap. However, we acknowledge that some people prefer to use the term 'sensibility to pain' on the grounds that it is more objective. Scientists are still striving to define and understand the broader concept of consciousness.

Where the process leaves the animal unconscious but alive, often referred to as 'simple stunning', there is a risk of consciousness returning. These methods must be followed by a 'killing method' that is completed before the animal regains sensibility. Other processes result in instantaneous death and are often referred to as 'stun-kill' methods. Some stunning methods vary in whether they achieve 'simple stunning' or 'stun-kill', depending on the parameters used, species and context.

Killing methods

'Simple stunning' needs to be followed by a killing method, performed before the animal regains consciousness. Killing methods cause irreversible damage to the medullary region of the brain responsible for breathing, and other parts of the brain responsible for consciousness, either destroying brain tissue through direct physical trauma or by depriving it of oxygen required for survival.

When animals are killed in slaughterhouses (see p. 265), hypoxia (a reduction in the oxygen supply to tissues) is generally induced by exsanguination (bleeding out) following the severing of the major blood vessels supplying oxygenated blood to the brain, typically the carotid arteries of the neck (Figure 17-1). This method is also required for meat hygiene reasons when animals are killed for human consumption. In mammals and birds, a knife is typically used to manually cut these vessels, and this is generally mechanised in large-scale

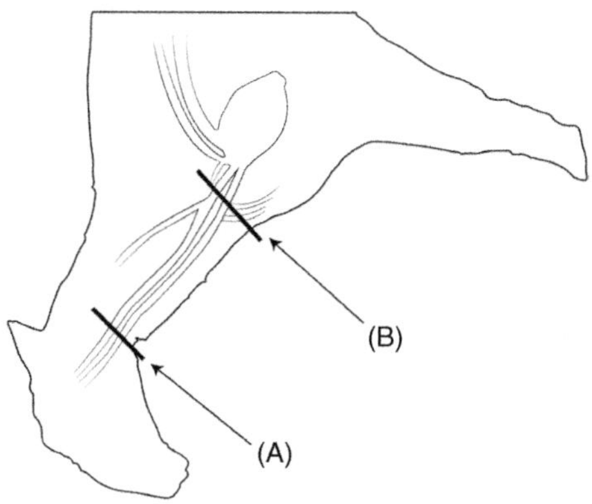

Figure 17-1. Bleeding out involves severing the carotid arteries (A) or the brachiocephalic trunk (B) where these vessels originate. Animals are typically hung upside down once unconscious to speed the rate of blood loss. Adapted from Humane Slaughter Association (HSA).[198]

poultry production. In cattle and pigs, a cut may instead be made at the thoracic inlet to sever the vessels emerging from the heart. In fish, the gills are cut. An alternative method for creating hypoxia is to induce ventricular fibrillation by applying electrical current to the heart so it can no longer pump blood. When a penetrative captive bolt is used, the hole created in the skull allows for a killing method called 'pithing' (see p. 260).

'Non-stun' killing

In various contexts globally, animals are killed by exsanguination without prior stunning for a range of reasons. In these situation the animals remain conscious for variable periods while they bleed out, resulting in associated welfare harms (see p. 274).

Stunning methods

There are four main categories of stunning methods:

- Mechanical.
- Electrical.
- Gaseous (controlled atmosphere stunning).
- Chemical (lethal injection/immersion).

Each is described in greater detail below.

Mechanical stunning

Mechanical stunning methods induce loss of consciousness by delivering force that directly traumatises the brain. They include captive bolt devices, firearms with free projectiles, manual percussive blows, maceration and neck (cervical) dislocation.

Captive bolt devices

Captive bolt devices deliver a forceful blow to the skull that immediately renders the animal unconscious, usually by causing rapid brain movement and pressure changes. Non-penetrative devices have a bolt with a broad end (Figure 17-2) that can severely damage the brain without entering the skull cavity, whereas penetrative devices additionally fire the bolt through the skull bone, penetrating and damaging the soft tissues of the cortex, thalamus and brainstem. They are generally 'simple stun' methods, though in certain animals, they cause enough damage to 'stun-kill'. Therefore, EU law mandates the use of a killing method 'without delay', with limited exceptions where devices are known to reliably 'stun-kill'.

Captive bolts are reliable methods when correctly applied by competent personnel. The amount of force applied is critical, being determined by the weight and velocity of the bolt, combined with the position and direction of the shot (Figure 17-3). Appropriate parameters differ among species and animal sizes, and the correct device must be used and properly maintained. Individual restraint is usually required to ensure an accurate shot.

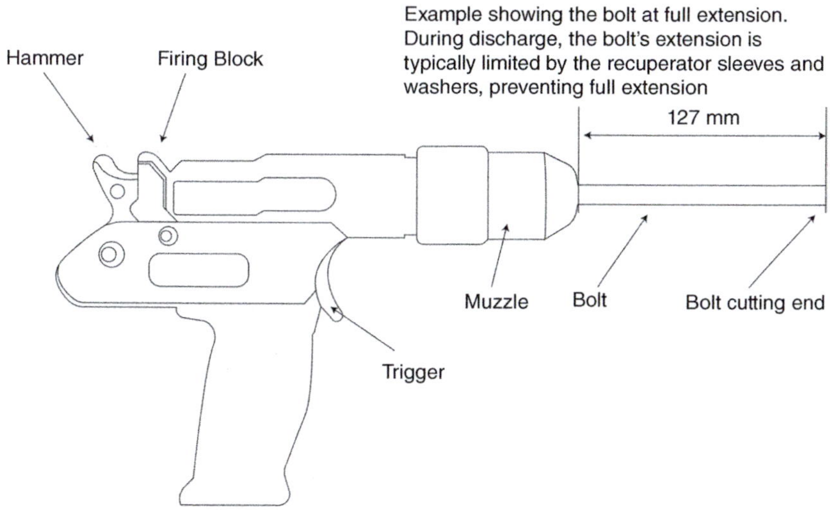

Figure 17-2. Penetrative captive bolt device. Adapted from technical materials available from www.accles-shelvoke.com.

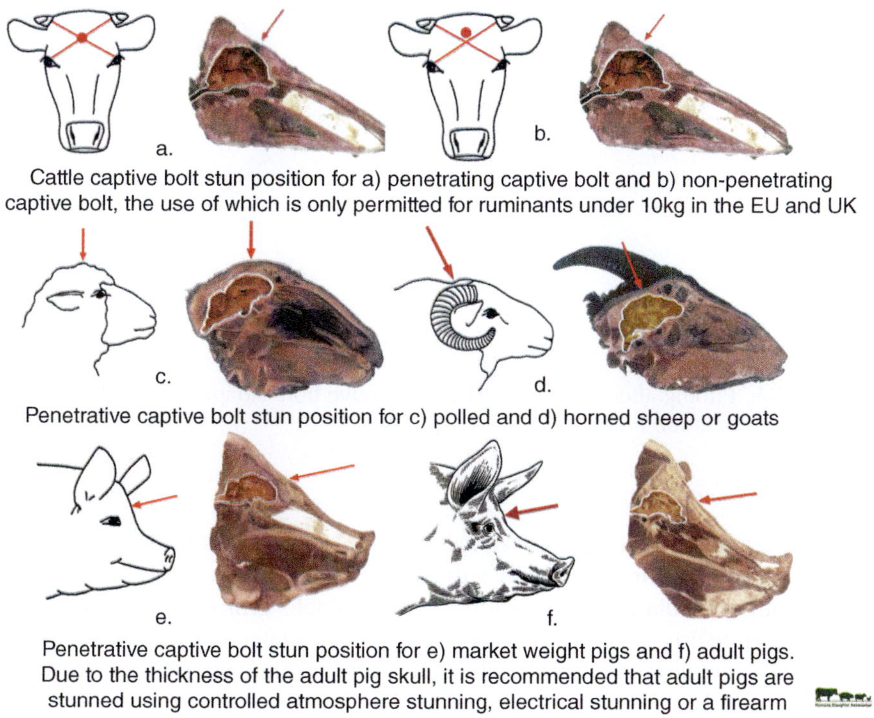

Figure 17-3. Stunning position for cattle, sheep and pigs. This is part of an extensive source of information available from the Humane Slaughter Association on humane slaughter and the killing of animals. Reproduced with permission from Humane Slaughter Association (HSA).[198]

The most common killing method following captive bolt stunning is exsanguination. This is required where animals are used for human consumption.

Following the use of a penetrative captive bolt, an animal can be then killed by pithing. In this method, a metal or plastic rod is inserted via the hole created by the bolt and manually manipulated to destroy the brain tissue and spinal cord (Figure 17-4). Pithing is also useful for reducing the extensive kicking action that follows stunning, but its use is often restricted. For example, pithing has been banned in EU countries in ruminants intended for human consumption due to the small risk of the meat being contaminated with the pathogens that cause transmissible spongiform encephalopathies. Pithing is useful when killing animals on-farm for disease control purposes, where bleeding animals and containing the blood could be difficult or a biosecurity risk.

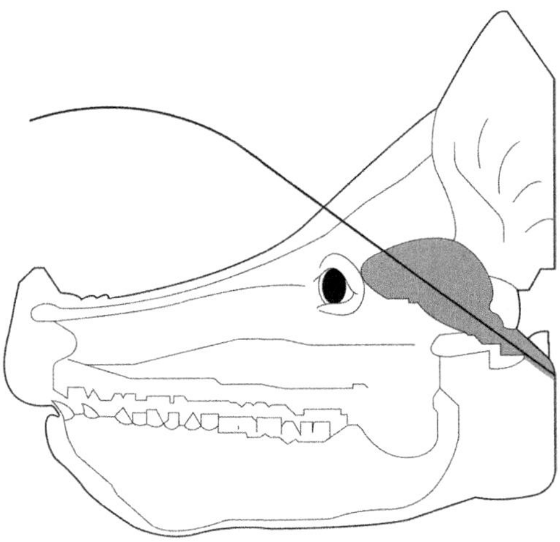

Figure 17-4. Pithing after using a penetrative captive bolt. Adapted from Humane Slaughter Association (HSA).[198]

Firearms with free projectiles

Firearms that use free projectiles include shotguns, rifles and pistols. Like captive bolts, these firearms apply sufficient force to the skull surface to make the animal immediately unconscious. It is a reliable 'stun-kill' method when an appropriate firearm is used correctly by competent personnel, as this causes extensive brain damage. The correct model, ammunition and positioning must be used, and there are additional health and safety considerations for operators. Under field conditions, very little restraint may be needed, resulting in less handling stress compared with animals forcefully restrained for close contact stunning and killing when using other methods.

Manual percussive blow

Manual percussive blows to the head can effectively 'simple stun' or 'stun-kill' neonates, other small animals and fish. However, they should not be used routinely in terrestrial animals because it is very difficult to consistently produce the correct force and positioning, and viable alternatives are available. Manual percussive blows are performed using a pole or hard object to strike the animal's head, or by swinging the animal to hit its head against a hard surface, such as a wall. In fish, a mallet-type object, called a 'priest', is often used. As a welfare safeguard, EU law limits the type of terrestrial species, weight (which must be under 5 kg), and the number of animals that can be stunned in this manner.[196] The laws of England and Wales further limit usage.[195,197]

Maceration

Maceration devices use rapidly rotating blades or rollers to crush tissues. EU and UK laws allow these to be used to 'stun-kill' chicks up to 72 hours of age and to destroy eggs containing embryos. Death is immediate when properly performed.

Neck dislocation

The neck dislocation method can be used in chickens and other poultry. It involves manually or mechanically stretching and twisting the neck aiming to dislocate the atlanto-occipital joint (between the skull and first cervical vertebra) or the joint between the first and second cervical vertebrae, thus severing the spinal cord and blood vessels in the neck and inducing cerebral ischaemia, as well as inducing concussion due to traction on the brain. This method requires minimal or no equipment and, therefore, is particularly useful in emergency situations when no other method is available. However, neck dislocation does not consistently render chickens immediately unconscious. EU law prohibits the routine use of this method and, for emergency situations, sets weight limits for birds. A mechanical method (using a device that increases leverage) may be used for poultry up to 5 kg, while a manual method (using only the hands) is permitted for poultry under 3 kg. In addition, there is a daily limit of 70 birds per person for manual methods.

Devices that crush the bird's neck cannot be used as stunning devices, because many birds will be paralysed by this method, remaining conscious and suffering for a period of time before dying from their injuries.

Electrical stunning

Electrical stunning methods pass electrical current through the brain tissue at a frequency, current and voltage that causes electrical brain activity analogous to that in humans undergoing grand mal seizures seen in some forms of epilepsy. In humans, there is overwhelming evidence that this is associated with unconsciousness, and the same is presumed true of animals. Birds and fish show different electrical brain activity and physical signs to mammals, so particular care is needed to ensure they are unconscious, not simply paralysed by the flow of electric current through their bodies. Laboratory testing using electroencephalography (EEG) or electrocorticography (ECoG) in a simulated commercial setting is important for verifying that electrical parameters are effective prior to commercial application.

There are various ways of applying electrical current.

'Head-only' electrical stunning

In 'head-only' stunning, the current passes through the head and brain only. It is a 'simple stun' method, producing a short period of reversible unconsciousness. To ensure there is no recovery of consciousness in slaughterhouses, animals are usually shackled, hoisted and bled out without delay so that hypoxia leads to brain death before they have had time to recover. 'Head-only' stunning is commonly used in slaughterhouses for sheep, goats, rabbits and pigs (Figure 17-5). A variation of this method, applies the electrical current in two phases. First, electrodes are placed across the head to induce unconsciousness. Then, electrodes are applied across the heart to cause cardiac arrest and death. This method is most commonly used in pigs and can be performed either manually or automatically. Automated systems are designed to handle the high processing rates required for pigs in many slaughterhouses.

In cattle, conventional 'head-only' stunning doesn't lead to a sufficiently long period of unconsciousness to ensure death by bleeding occurs before consciousness is regained. Therefore, where used, it is immediately followed by an automated process that induces cardiac arrest to ensure death. An electrical stunning method for cattle, using 'Single Pulse Ultra Current (SPUC)', is being developed, which produces longer-lasting unconsciousness than conventional 'head-only' stunning techniques. It should be particularly useful for Halal-compliant processing, by keeping cattle unconscious for the time it takes them to die from exsanguination; if it is able to made commercially viable.

Figure 17-5. Head-only stunning and head-to-body stunning of sheep. Adapted from Humane Slaughter Association (HSA).[198]

In the UK, 'head-only' electrical stunning is commonly used for the small-scale killing of chickens and other poultry on farms for direct supply to consumers. To ensure the procedure is humane, the equipment must be sufficiently powerful, properly maintained, and used correctly so that enough electrical current passes through the brain to effectively stun animals and for sufficient time until the brain is irreversibly damaged by bleeding out (see p. 257). Welfare risks may be increased by the significantly lower level of regulatory oversight on farms compared with slaughterhouses.

'Head-to-body' electrical stunning

'Head-to-body' electrical stunning delivers one cycle of electrical current, which flows through the heart and brain simultaneously (Figure 17-5). Proper electrode placement and parameters are crucial to ensure cardiac arrest occurs only in the unconscious animal, as cardiac arrest when conscious would likely cause extreme pain.

Waterbath stunning is a specific method of 'head-to-body' stunning used in chickens and other poultry. The birds are inverted, with the shanks of their legs placed in metal shackles that form one part of an electric circuit. Their heads make contact with a water bath connected to the other part of the circuit, allowing an electric current to pass from head to feet. This produces a whole-body stun that simultaneously affects both the brain and the heart. These systems are typically automated and function on a large scale. Where low-frequency currents are used (50 Hz), this method can 'stun-kill'. However, at high frequency currents, it is a 'simple stun' and must be followed by a killing method, usually bleeding after cutting the carotid arteries. Some chickens risk being immobile due to paralysis, rather than unconsciousness, if the correct evidence-based parameters are not used, and there are additional welfare harms associated with the handling and inversion of birds. Water-bath stunning has largely been replaced by carbon dioxide (CO_2) gas stunning in the UK (see below).

However, water-bath stunning is still used in small slaughterhouses where automated gas systems are impractical. Also, higher frequency reversible water-bath stunning remains popular for Halal slaughter, as religious text interpretations require the animal to be killed by the throat cut and exsanguination only, and not by other methods.

Controlled atmosphere stunning (gaseous and low atmospheric pressure)

In the UK and EU, controlled atmosphere stunning (CAS) is commonly used for pigs and poultry, but it is not generally permitted for ruminants. Unlike electrical and captive bolt stunning, CAS systems do not render animals immediately unconscious; therefore, they may experience some degree of suffering before becoming unconscious, depending on the system used. The main methods are carbon dioxide, inert gases and low atmospheric pressure stunning (LAPS).

Carbon dioxide (CO_2)

In the UK, the vast majority of pigs and poultry are 'stun-killed' using CO_2. Exposure to high concentrations of CO_2 displaces oxygen and other gases in the air, leading to unconsciousness and then death as a result of hypercapnia and hypoxia.

Carbon dioxide concentrations used vary depending on the species and context. The concentration of CO_2 needed to kill is 40% or greater. However, inhaling these concentrations is painful, and as CO_2 forms carbonic acid in the respiratory tract, it irritates the airways and causes distress. It is routine practice in slaughterhouses to expose pigs to CO_2 concentrations of 90%, even though the pigs typically undergo 15–30 seconds of aversive induction before they are unconscious.

In chickens, a much less aversive multiphase method is used. The bird initially inhales up to 20% CO_2, which does not cause pain but renders the bird unconscious before the levels are increased to the 40% concentration that kills, but would be aversive in a conscious bird. Animals convulse when stunned, but this is thought to occur after they have lost consciousness. However, the adverse welfare effects of exposing poultry to levels of CO_2 before unconsciousness occurs are currently being investigated, as hyperventilation and other signs indicating discomfort are observed.

Inert gases

Inert gases, principally argon and nitrogen, work by displacing the oxygen fraction of the air, causing hypoxia. Unlike CO_2, they do not have any additional acidifying and sedative effects. The duration of anoxia determines whether the method is reversible, that is, produces a 'simple stun' or a 'stun-kill'.

Inert gases are much less aversive than CO_2. They can be used alone or in combination with CO_2 to stun animals. However, commercial CAS methods using inert gases are not yet well developed in the UK or EU, though they are generally now considered a more humane alternative to CO_2 in pigs. Currently, their use for stunning is limited to containerised gas units (CGUs) for poultry, primarily used for large-scale culling on farms for disease control or other management reasons, as well as for routine culling of day-old male and non-viable chicks in hatcheries.

Low atmospheric pressure stunning (LAPS)

Low atmospheric pressure stunning (LAPS) is a 'stun-kill' method that is only legally permitted in the UK and EU for poultry. Birds typically remain in their modular crates and enter a device in which oxygen is removed, resulting in conditions similar to those found at high altitude. i.e. gradual atmospheric decompression as the available oxygen is reduced to less than 5%. Hypoxic conditions cause loss of consciousness and death with minimal distress.

Chemical stunning

An overdose of anaesthetic (for example, pentobarbital) directly into a vein is frequently used in companion animals and horses to kill them humanely. It causes minimal pain, limited to the needle entering the skin and some stinging as the initial dose enters. The loss of consciousness and sensibility is similar to that experienced when being anaesthetised. Minimising restraint stress is key to welfare, so this method is more suitable for some animals and contexts than others.

Other methods of administering an anaesthetic overdose, such as direct intracardiac injections, can cause considerable pain. Such methods should be preceded by deep sedation to make the animal first insensible to pain. Fish can be killed humanely by immersing them in and overdosing them with liquid anaesthetic agents.

Chemical stunning of fish and livestock may be used when they are not intended for human consumption, such as for disease control purposes.

Signs of an effective stun

An effective stun makes an animal unconscious, and therefore, insensible to pain, distress and any other adverse welfare state. In laboratory settings, brain activity can be measured and monitored objectively using an electroencephalogram (EEG) or electrocorticogram (ECoG), with a lack of electrical activity or specific wave patterns indicating the animal is unconscious. This monitoring is not practical in field conditions. However, the unconscious state is associated with certain behavioural indicators and lost reflexes, indicating that the brain stem is not functioning. Such behaviours serve as proxy measures for assessing unconsciousness and insensibility in field conditions. EFSA has evaluated these behaviours in commonly farmed species and used the knowledge to produce guidelines suitable for monitoring signs of unconsciousness in a commercial setting. They are crucial for the early detection of any return of consciousness, where immediate re-stunning is vital for protecting welfare.

Slaughterhouses

Most humane killing of farm animals in the EU occurs in specially designed facilities known as slaughterhouses or abattoirs. These facilities come in various designs, depending on the species and methods of killing. After transporting the animals from the farm, they are typically unloaded and held in a lairage before being moved to a designated area where they will be restrained, stunned and killed. After this process, the meat is prepared for consumers from what is now a carcass.

Animals are removed from vehicles as soon as possible after arrival, ideally immediately. In EU countries and the UK, chickens, other poultry and rabbits are transported in crates. Arrival is timed to ensure immediate movement for killing; however, if there is a delay, crates containing animals should be kept in temperature-controlled lairage sheds after being removed from the vehicle.

Transport

Animals are typically transported by road, but many are also transported by sea and rail. Significant advancements in vehicle design mean that specialised vehicles are now available to better meet the animals' needs, including providing water and controlling their microclimate during transit. Figure 17-6 highlights some vehicle design features that improve welfare.

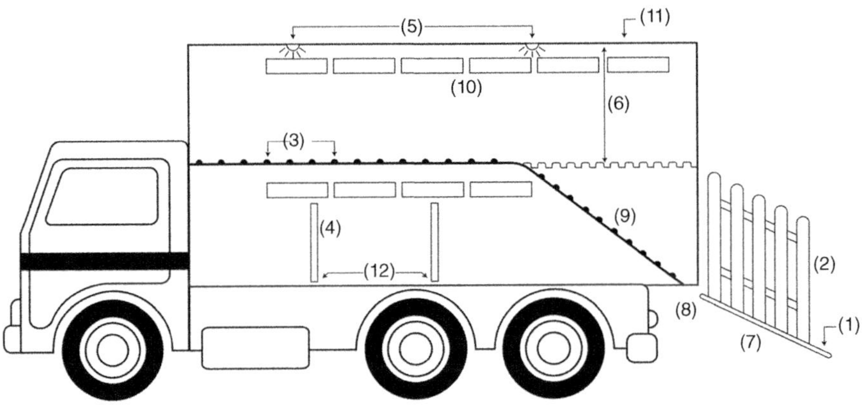

Figure 17-6. Examples of vehicle design features that improve welfare. (1) Loading ramp with foot battens. (2) Sidegates. (3) Foot battens and/or non-slip flooring on all floors. (4) Internal partitions. (5) Adequate lighting for stock inspection. (6) Height between floors must allow animals to stand naturally and provide adequate ventilation. (7) Slope of loading ramp (the maximum angle is mandated by law in the EU and UK). (8) The step at the top of the ramp is minimised. (9) Slope of internal ramp minimised. (10) Adequate ventilation on all floors. (11) Roof to protect from the weather. (12) No large gaps between internal partitions and floors. Adapted from Humane Slaughter Association (HSA).[198]

Common welfare hazards

- Handling quality.
- Assessment of fitness for transport.
- Road conditions, travel time and quality of driving.
- Vehicle and facility design.
- Space allowances during transportation and lairage.
- Mixing unfamiliar animals.
- Exposure to unfamiliar environments, situations and stimuli.
- Methods of restraint prior to stunning or killing.
- Stunning methods and equipment maintenance.
- Use of non-stun slaughter methods.

Welfare hazards are the resources (inputs) that can lead to poor welfare outcomes. The likelihood of poor welfare outcomes and their severity (i.e. the risks) can vary. Severity encompasses both duration and intensity. Common welfare outcomes are identified below, structured by the Five Freedoms framework.

Welfare is determined by how well animals can adapt, both physically and mentally, to their environment (see p. 8). Relevant descriptions of 'free-living' animals are provided throughout this book to highlight the adjustments they must make while being handled, transported and killed. Here, 'free-living' is defined as animals living freely in an environment with the resources for which they have evolved and making their own choices. However, such comparisons provide only part of the welfare picture; welfare science (see Chapter 2) is essential for a proper understanding of the welfare impact of different operational systems.

Key determinants of welfare are the knowledge, skill and motivations of personnel, as well as whether they have sufficient resources (see pp. 24–25). Where these factors are optimal, welfare will be as good as it can be within a given operational system. However, there are often persistent welfare issues that are challenging to resolve without altering the operational system itself.

Common welfare outcomes

F1. Freedom from hunger and thirst

Animals experience hunger or thirst when unable to eat and drink for extended periods.

Journey length

When transported, the risks of experiencing hunger and thirst increase with longer journeys. Therefore, it is preferable to use slaughterhouses located close to their farm of origin.

Vehicle design

Some vehicles transporting cattle, sheep and pigs are fitted with water drinkers, but these are no longer thought to mitigate thirst as intended, and animals may still be thirsty once they are 8–9 hours into the journey, if not before. If vehicles lack drinkers or the design is incorrect for the species, such as using pig drinkers for calves, thirst will likely occur sooner. The comfort level during transport also affects thirst, with factors such as heat stress exacerbating it.

Feeding animals during transport is very challenging. It is impossible when chickens and rabbits are confined in crates. Also, stopping en route may not help if animals are reluctant to eat.

Contingency plans and scheduling

To mitigate hunger and thirst, vehicle arrivals must be appropriately scheduled to prevent queues from forming at the slaughterhouse. Drivers should also have contingency plans in place to protect welfare during unexpected journey delays.

On arrival at the slaughterhouse, cattle, sheep and pigs should be promptly unloaded, preferably within 15–60 minutes. They should be placed in lairage pens with sufficient water drinkers for the group and provided with feed as appropriate, depending on when they will be killed. Chickens and rabbits remain in their crates until killed, which prevents them from being fed or watered, but crates should be unloaded into custom-designed, climate-controlled areas of the lairage to keep them as comfortable as possible.

Withholding food

In some species, such as chickens and pigs, hunger may be exacerbated by withholding food on the farm before transport. This is a planned measure in poultry and pigs, used to reduce gut contents, thereby minimising carcass contamination during processing. In pigs, this practice also limits their nausea and vomiting. In farmed fish, feed is routinely withheld (see Chapter 14) to try to maintain water quality when they are 'crowded' for transport and handling.

F2. Freedom from discomfort

Multiple, interrelated factors influence an animal's comfort level during vehicle transport, including temperature, air quality, movement, sounds, floor surfaces and stocking density.

Vehicle and crate design

Vehicle design, combined with the actions of handlers and drivers, is key. Vehicles should have sufficient headspace for animals that are standing. Crate dimensions should allow chickens and rabbits to sit naturally without their heads, combs or ears touching the crate roof, and dimensions should provide sufficient space for all chickens or rabbits to lie comfortably and adjust their position. Animals that walk onto the vehicle require sufficient space and a stable floor grip to adjust their posture and balance in response to the vehicle's movements. Bedding may provide additional grip or comfort if animals lie down, something they may choose to do on long journeys.

Thermal comfort

To mitigate discomfort, drivers must understand and monitor how the microclimate around the animals varies and take action based on this information. For instance, combined temperature and humidity levels determine an animal's risk of heat or cold stress, and these parameters differ significantly according to vehicle design and usage, including space per animal, deck height, compartment construction, whether the vehicle is static or moving and whether the side flaps are raised or closed. Some parts of the vehicle pose greater risks than others. Heat stress poses a greater risk than cold stress for most livestock species. However, 'end-of-lay' hens are particularly vulnerable to cold due to their poor feathering and low body fat content and the open-sided vehicles used to transport them. Ideally, drivers will have access to mechanical ventilation and air conditioning to better control conditions around animals.

Route length, road quality and driving style

Route length, road quality and driving style are important factors. A longer journey on good quality, straight roads may serve animal welfare better than a shorter journey on poor quality, winding roads. Scheduled vehicle stops during long journeys often do not allow animals to rest as intended.[144,199] To properly rest, they need to be unloaded into suitable facilities.

Lairage conditions

On arrival at the slaughterhouse, lairage conditions and length of time spent there are important determinants of comfort. Animals waiting on vehicles and those unloaded into pens need sufficient shelter, shade and quiet surroundings. Using fans can benefit poultry and rabbits held in crates in hot weather. Lairage pens should provide clean, dry, comfortable resting areas for all animals. Some animals may need specific attention. For instance, milking cows must be either milked or slaughtered before the udder becomes distended. Where farmed fish are 'crowded' for handling and transport to live fish processing plants (i.e. dedicated fish-specific slaughterhouses), they may experience discomfort due to reduced water quality (see Chapter 14).

Prior learning

Prior experiences can affect discomfort levels. For instance, animals raised extensively outdoors may be less accustomed to the different types of flooring and noises during transport and lairage, making them more uncomfortable than those raised indoors.

Restraint

When captive bolt devices or electrical stunning are used, physical restraint is generally used to ensure precise placement of the equipment and the delivery of an effective stun. However, such restraint—particularly when it involves restricting movement or fixing the head—can cause discomfort and distress. It is therefore crucial that restraint methods are carefully designed, applied, and monitored to minimise any negative welfare impacts.

Factors that cause discomfort can escalate or combine, resulting in pain, injury, disease, fear and distress.

F3. Freedom from pain, injury and disease

Common sources of pain and injury during handling and transport are falls, collisions with objects and other animals, and fights when unfamiliar animals are mixed. Pain and injury are likely associated with fear and distress. In turn, fear and distress often make animals flightier and increase the risks of pain and injury. The actions of personnel are key and, ideally, will be shaped by an understanding of behaviour and welfare (see F5). Well-designed facilities should assist them.

Handling quality

As already described (see F2), animals need to be able to balance during journeys and to rest if fatigued. Overcrowding and poor driving or road conditions can cause animals to fall. If unfamiliar animals are mixed, fighting usually escalates rather than settles during the journey. Horned animals can seriously harm those without horns, and the two should generally not be mixed.

Chickens and rabbits are loaded into crates on their farm of origin, prior to transport. Some are later unloaded from crates at the slaughterhouse, depending on the stunning and killing methods used. These species have the same welfare needs to be protected from pain and injury as the larger livestock species. However, their small size and sheer volume make it more likely their needs will be disregarded, particularly when handlers are operating under time constraints. Proper supervision is necessary to ensure that animals are carried with their body weight properly supported, ideally not suspended upside down by their limbs, as is common industry practice for some poultry, nor ever lifted solely by their wings, ears, tails or other body parts.

Large animals, such as cattle, demand more consideration from handlers for safety reasons as they can easily harm themselves and personnel when fearful or agitated. Their emotional state is generally easier to 'read'. However, larger animals can still be overpowered as handlers tend to have handling 'aids' that they can use inappropriately. This can be a particular problem in pigs as they are less easily controlled using flight distances than ruminants and therefore more likely to suffer pain or injury during handling.

Fitness for transport

Some animals do not have the fitness levels to cope with transport and killing operations and should be killed on the farm. EU laws cover this in detail and are similar to WOAH guidance[9] as to which animals are unfit for transport, including those unable to move without pain or walk unassisted, those with severe open wounds or prolapses, pregnant animals in the last 10% of gestation, those that have given birth the previous week and newborn animals in which the navel has not healed. However, owners may still be tempted to transport unfit animals due to commercial pressures. If an animal is killed in a slaughterhouse, then the meat can be salvaged for human consumption, but if killed on a farm, there are costs associated with killing and subsequent carcass disposal. Additionally, it is often considered accepted practice to routinely transport some animals at the end of their productive life, where various age-related pathologies can increase their suffering. Examples are 'cull animals', such as 'end-of-lay' hens with keel and limb bone fractures or breeding sows having difficulty walking due to painful limb and foot pathologies.

Economics

Pain and injury before killing carry economic, as well as welfare, costs. For instance, meat from bruised, fatigued or distressed animals is often less valuable. Examples include 'pale soft exudative' meat from pigs, 'dry firm dark' meat from cattle and 'gaping' flesh from fish. However, some animals at the end of their productive life are of such low value that there is no financial incentive to improve conditions, and lower-quality meat is simply sold at a lower but still overall profitable price.

Facility design

There must be non-slip flooring at all stages, as handling and transporting animals humanely on slippery surfaces is impossible. To help prevent bruising, gates, fences and chutes should not have projections or other sharp surfaces that could cause injury. Vertical slide gates should be counterweighted and, if possible, padded to avoid causing back bruises. Loading and unloading are the most stressful parts of vehicle transport, requiring animals to have considerable stamina and fitness. To optimise welfare, loading bays can be designed so that ramps have gentle inclines, requiring less exertion, and side gates to prevent animals from falling off.

Stunning and killing methods

Good operator skills, knowledge, attitudes and the usage of appropriate, well-maintained equipment can optimise welfare during any form of stunning and killing.

Operators must be able to recognise the signs of effective stunning when using different methods and in different species. Animals that are stunned must remain unconscious until they are killed by a follow-up method if required, usually bleeding out via a deep neck cut that severs vessels to the brain in mammals and birds, or cutting the gills in fish. Logistics may limit what the operator can achieve. Ideally, they would check every single animal to determine whether it is effectively stunned and killed. However, the high killing rates of rabbits, chickens and fish make it almost impossible to do this individually for these species. Instead, a compromise is made based on the reliability of these stunning methods by assessing a sample of animals directly, while the rest are presumed to be properly stunned based on monitoring inputs, such as the electrical current recorded on control panels.

Strengths and limitations of different killing methods

It can be difficult to achieve the ideal restraint, stun and kill for every animal. When using different methods, there are various pros and cons to weigh up.

Some methods can immediately and effectively stun an animal, but the necessary handling and restraint to apply the method may upset them. For example, it is a highly skilled task to handle a small group of pigs in a 'crowd pen' and keep them all calm while electrically stunning each individual in turn by applying metal tongs across their head.

Some methods pose little restraint stress, but the stun is not immediate. For example, controlled atmosphere stunning (CAS) in pigs, using CO_2, causes little restraint stress as pigs are managed in small groups and often with automated barriers. Although this method reliably stuns animals, it is highly aversive for at least 15–30 seconds. The inhaled CO_2 creates acidic conditions that irritate the mucous membranes and cause pain. Pigs are also likely to feel short of breath. In contrast, inert gases such as argon are not considered aversive, but the associated hyperventilation and potential feelings of breathlessness are still being evaluated. Production advantages include a higher throughput of animals than if they are stunned individually and a lower level of meat degradation.

Many chickens are now stunned using CAS systems that gradually increase the concentration of CO_2 in a manner that ensures the birds are unconscious before the gas becomes highly aversive. They typically remain in crates during the stunning process and are only taken out once they are either stunned or dead. This minimizes the stress related to handling. However, the welfare impact of inducing unconsciousness, especially if prolonged, has yet to be fully evaluated.

Some methods do not always reliably stun the animal, and the restraint is distressing. Alternatives are needed. An example is the electrical stunning of chickens using a water bath, particularly when the electrical parameters are not set appropriately. Handlers remove birds from the transport crates and then suspend them upside down by their legs, which are pushed into metal 'shackles', causing pain, fear and distress. Refinements, such as using body or breast supports to support the weight of the shackled birds until they are stunned, can reduce but not remove their pain and distress. Some birds experience painful electric shocks if their wingtips accidentally touch the water before their head. For some, if electrical parameters are not set appropriately, insufficient electrical current is delivered to the brain to cause unconsciousness, and they risk being bled out whilst still conscious.

Some animals are hard to stun and kill humanely due to logistical challenges. For instance, the vast numbers of fish that are caught and landed on the boat deck simultaneously in commercial capture fisheries have already suffered considerable trauma (see Chapter 14). Many are then killed using methods that cause pain and distress, such as being placed in ice slurry where they lose consciousness slowly or having their gills cut whilst fully conscious. In contrast, humane electrical and mechanical percussive stunning methods have been developed for some farmed fish. They are routinely used on commercial Atlantic salmon farms and some trout farms.

Comparative contexts of emergency and humane killing

In some situations, despite careful planning, circumstances may arise where an animal should be killed immediately to relieve its suffering, even if not all legal requirements, such as using the ideal stunning equipment, can be met. In such cases, it may be in the animal's best interest to be humanely killed without delay rather than waiting to fully comply with all legal provisions, if the animal will suffer less as a result. Accordingly, EU and UK legislation recognises that in cases of 'emergency killing', certain legal requirements may not be fully met where this is necessary in the animal's best interest. For example, if a chicken is experiencing severe pain and suffering, it may be justified to kill the bird using neck dislocation to prevent further suffering if there is no immediate alternative method available. This method involves manual stretching and dislocation of the neck, typically performed on an inverted bird. However, loss of consciousness is not always immediate; therefore, the welfare outcome may be suboptimal (see p. 261). Far preferable is ensuring that effective planning has been undertaken and that appropriate resources are available, so that more reliable methods – such as the use of a captive bolt device capable of delivering a consistent and effective stun without the need to invert the bird – can be used instead.

When using firearms with free projectiles, shots should be accurately aimed at the head to ensure instantaneous loss of consciousness and death. Other shot placements, such as targeting the heart or neck, may result in poorer welfare outcomes if unconsciousness is not immediate. However, the latter approach may still be justified on welfare grounds in situations where animals cannot be safely approached or handled at close range, and a head shot cannot be reliably performed.

In some low-income regions, manual percussive stunning with a heavy mallet or poleaxe though far from ideal may be the most practical option available. In such contexts, it may be preferable to use this method rather than to introduce complex equipment that is difficult to maintain, operate, or apply reliably. Nevertheless, welfare improvements can often be achieved through other parts of the killing process, such as improving handling, transportation, and lairage practices, supported by targeted training and modest resource investment, rather than relying solely on equipment upgrades.

Sentience in foetuses and embryos

The sentience of early developmental stages needs to be considered, for example, when killing pregnant animals. The neuroanatomy of most mammalian foetuses is probably not well enough developed for them to perceive pain until the last third of gestation, and even at this stage, low oxygen levels are presumed to keep the foetus in a sleep-like state for much of the time. EFSA recommends not opening the uterus of a dead animal until at least 30 minutes after the mother has died, as by this time, the foetus will be dead and does not risk becoming conscious through gasping air (see p. 30).[19]

'Non-stun' killing

'Non-stun' killing, where animals are killed by bleeding while still conscious, is performed for various reasons in different countries, including economics, lack of welfare knowledge or concern, lack of resources and laws limiting the use of firearms, as well as meeting the religious requirements of Halal meat for Muslim consumers and Kosher meat for Jewish consumers. Non-stun killing causes varying degrees of pain, injury, fear and distress as animals experience the deep neck cut, the sudden drop in blood pressure as they haemorrhage, and suboptimal restraint methods may also cause distress. Non-stunned animals can remain conscious for variable amounts of time during exsanguination. There are various interpretations of relevant religious texts. Reversible forms of stunning ("simple stunning," as opposed to "stun-kill"), such as electrical head-only stunning of small ruminants and poultry, may be acceptable in some Halal systems, provided the animal remains alive at the time of the cut. In contrast stunning of any sort is generally not permitted in shechita.

F4. Freedom from fear and distress

Fear and distress often accompany pain and injury. They can also occur independently. Most animals find sudden exposure to new environments, people and animals threatening. When handled, transported and killed, they are exposed to unfamiliar sensations such as vibrations, moving floors, engine smells and sounds. The extent to which animals feel fear depends on their species, breed, age, sex, prior experience and learning. Many farm animals can become habituated to new experiences if given the chance to learn and form positive associations. For example, a domestic horse or farm animal that has frequently travelled to shows may not view travelling to a slaughterhouse any differently. However, few farmed animals have this opportunity to become used to travelling.

Handlers need to recognise signs of fear and distress in different species. For instance, cattle and pigs tend to vocalise, whereas sheep often remain quiet, meaning their distress is more easily overlooked. Fearful animals are generally hyperalert and easily agitated, which makes accurate stunning more difficult. For instance, fearful pigs may pile on top of each other in a 'crowd pen', and fearful cattle move their heads around in the stun box. Employing best practices based on understanding the animal's normal behaviour (see F5) can mitigate fear and distress.

In many cases, it may be better to keep social animals in close contact with each other rather than isolate them to kill them. UK laws used to require ruminants and pigs to be killed out of sight of others. However, following research, it appears that this is not necessary if the animals are killed without distress, and the law was changed accordingly. In the UK, it is still illegal to kill horses in sight of one another. However, the Animal Welfare Committee recommends that when it comes to killing feral or nervous animals, it can benefit their welfare if they are with another familiar animal rather than being isolated.[200] This recommendation presumes that the person performing the procedure is skilled in safely managing horses and safeguarding their welfare.

F5. Freedom to express normal behaviour

Normal behaviours that contribute to good welfare, such as resting, eating and friendly social interactions, are severely restricted during transport and killing-related operations. Animals are likely to show more behaviours associated with lower welfare, such as greater flightiness and escape attempts linked to fear. Various strategies can help mitigate the welfare harms.

Using animal behaviour knowledge to improve welfare

All personnel could improve their knowledge of animal welfare and behaviour and practically apply this, choosing the most appropriate handling methods that minimise the animal's perceptions of restraint while still doing the job safely, efficiently and effectively.[194] People must understand how different animals perceive their surroundings and why they react in certain ways. They must also be skilled in assessing how the animals feel by observing their body language and altering their actions accordingly.

Consider cattle, sheep and pigs, all species that are moved on foot onto and off vehicles and through various parts of the slaughterhouse. They are prey species, having a flight zone that the handler must be mindful of. The flight zone is the area around the animal, which, if entered, causes the animal to move away. The size of this zone varies considerably among animals and according to the situation. For instance, it is smaller when the handler is calm and confident and tends to increase when animals become more vigilant in novel environments. If someone penetrates the flight zone too deeply or quickly, animals may panic and attempt to escape. Knowledge of the flight zone can be used to move groups of animals with minimal upset. For instance, to move cattle forward, the handler can step inside the flight zone, just behind the 'point of balance', located to the animal's side and just behind its shoulder (Figure 17-7). To stop them, the handler can step back outside. A common mistake is for handlers to position themselves at the head of the animal whilst prodding its rear when trying to move the animal forward through a raceway.

Given that cattle, pigs, and sheep usually live in stable groups and form social hierarchies, unfamiliar animals should not be mixed, or there will be fighting. Overcrowding of pens and other areas increases competition for resources and also causes aggression.

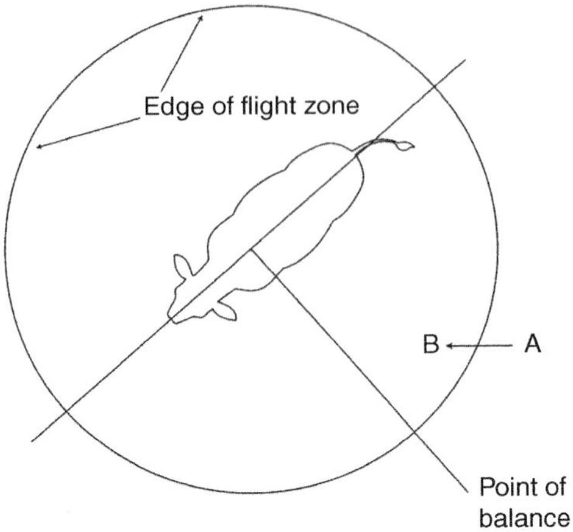

Figure 17-7. The flight zone of cattle and how it can be used to elicit movement. When the flight zone is entered – e.g. moving from point A to point B – the animal will move forward if the handler is behind the point of balance (typically at the shoulder), and backwards if the handler is in front of it.

The very different senses of livestock mean that things that are seemingly innocuous to people can alarm them. When animals refuse to move as required, there is often a simple solution if looked for from first principles. For instance, cattle and sheep have panoramic, wide-angle vision and readily detect the movement of distant objects because they have evolved to detect potential predators on the horizon while grazing. Therefore, they are easily startled by sudden movements. They have difficulty focusing on objects close by, especially immediately in front of them, and need to stop and lower their heads to assess these. They have a 'blind spot' immediately behind them. Pigs are adapted for woody undergrowth and have poor long-distance vision. Instead, they tend to use their sensitive nose and sense of smell to investigate their surroundings, which means they do not like to be driven forward. The hooves of different livestock are designed to grip specific terrains. Most livestock species are startled by sudden and loud noises, such as metal clanging, and can detect many noises inaudible to humans. They particularly dislike hissing noises. Their sense of smell is far superior to that of people.

Removing or minimising the things likely to distract the animals is paramount because they are moving through completely new facilities without prior training. Among the many possibilities that will cause baulking are sudden changes in floor type, drain covers, certain air drafts, clothing hung on a fence, pieces of plastic moving, fan blade movements and small objects left lying on the floor. Any one of these is enough to stop animals from moving forward. Handlers need to be patient and allow animals to walk at a natural pace, stopping along the route to allow them to investigate. Keeping animals calm could help prevent them from excreting pheromones that signal danger to other individuals.

Facility design

Knowledge of behaviour and welfare should underpin the design of facilities.

Remaining with other group members is very important to these social species. Cattle and sheep like to be able to see each other at all times. Pigs are less dependent on vision but still like close contact with other group members, communicating vocally and often maintaining body contact. Isolation causes fear and panic. Therefore, handling facilities should allow the group to move together, only placing them in a single file or isolating them at the latest possible moment.

Attention to the shape of races can be helpful. For instance, gentle curves can work well for cattle and sheep, taking advantage of their curiosity and willingness to follow the animals ahead. In contrast, sharp bends and corners should be avoided. The crowd pen leading to a single-file race should not be funnel-shaped, as animals risk being crushed through their desire to stay together. Instead, the opening to the final chute should be just wide enough for one animal to enter at a time. When single animals enter a stun box, they should be immediately stunned to limit their experience of isolation.

One key welfare advantage of commercial CO_2 gas stunning systems for pigs is that they allow the animals to be maintained in groups throughout the handling and stunning process. This is in contrast to automated electrical stunning systems, where pigs must move into a single file, onto restraint conveyors, to be individually stunned. Here, some pigs may resist moving forward, and as a result, an electrical prod is often applied to their hindquarters to encourage them to move, something they find aversive.

Solid sides to raceways will prevent animals from seeing fear-inducing distractions such as moving people, vehicles, other machinery and unfamiliar animals. Sounds should be dampened wherever possible.

Facilities should be designed so that animals cannot see when they are at a height, for example, by ensuring that conveyor restrainers have false floors and providing solid side gates on vehicle loading ramps. It is vital that floors have ample grip and that individuals are allowed to walk at their natural pace without being rushed. This makes animals feel more secure, and they are much less likely to fall. Animals cannot be moved humanely on slippery floors. Note that animals require considerable agility and stamina to enter and alight from vehicles. Loading ramps should be as level as possible, and foot battens for grip may be needed on ramps.

Welfare assessment protocols

Animal welfare assessment protocols currently available and under development have adapted a combination of resource-based and animal-based measures to the context of humane killing.

The Animal Welfare During Slaughter section of the WOAH Terrestrial Animal Health Code[9] provides a useful overview of various resource and animal-based measures that can be used during killing and related operations.

Dr Temple Grandin's work[201] has significantly shaped the approach to assessing welfare during transport and slaughter in the USA and other countries, offering detailed measures and audit guidance.[202]

The European Commission has published practical fact sheets[203] to support best practice, covering how to handle, restrain, stun and stun-kill various livestock species. They have also produced various Animal Transport Guides for the main livestock species.[204] The European Reference Centres for Animal Welfare will also provide practical advice on handling, transporting and humanely killing various livestock species.

Animal-based measures can also be used post-mortem, during processing, to retrospectively provide information about the welfare state of the animals on their farm of origin. This information can be used to improve the welfare of later generations of animals on those farms. Examples include the prevalence and severity of hock lesions and pododermatitis in meat chickens, indicating on-farm management issues that need to be addressed.

Advocacy examples

Reports and guidelines

There are credible reports and guidelines produced by official governmental bodies, veterinary professional associations, charities, individuals with special expertise and equipment manufacturers that can be used to advocate for good practice.

- The Humane Slaughter Association (HSA), a UK-based charity, offers extensive practical online guides covering the transport, handling and humane killing of commonly farmed livestock species.[198]
- The UK-based AWC Welfare at Killing Committee[205] has covered various topics, recently the use of nitrogen foam for culling poultry, preslaughter handling of horses, chick culling alternatives, emergency culling and depopulation of poultry affected by avian influenza, killing of farmed fish, and different methods of catching, carrying, collecting and loading poultry.
- In the EU, the EFSA has produced extensive reports covering the humane killing of a variety of commonly farmed species, which can be accessed online at their Welfare at Slaughter webpage.[206] There are also reports covering welfare during transport for small ruminants, equidae, cattle, pigs, domestic birds and rabbits.
- WOAH have published welfare recommendations in the Terrestrial Animal Health Code[9] that member countries are expected to follow, including

when transporting animals by sea, land and air, during slaughter and during killing for disease control purposes.[9]

- The BVA has set out its position on the welfare of animals at slaughter, with clear and extensive recommendations on how improvements can be made.[207] The American Veterinary Medical Association (AVMA) has produced guidelines for euthanasia, slaughter or depopulation, focused on the US context and legal framework.[208]
- Some equipment manufacturers provide helpful guidance. It is a legal requirement in the UK and EU countries that stunning equipment is sold with appropriate instructions concerning its use in order to promote better welfare. Some manufacturers, such as Frontmatec (Accles and Shelvoke) and BOCK Industries, collaborate extensively with users and provide additional information and videos on their websites.
- The World Society for the Protection of Animals (WSPA), now known as World Animal Protection, initiated the STEP (Slaughter Training and Education Programme) to enhance animal welfare during slaughter in countries with significant meat production. This programme primarily targeted China and Brazil, aiming to implement humane slaughterhouse practices and improve the overall treatment of animals. Training videos were produced on the slaughter of chickens, pigs and cattle for this international audience and are available via this link to WARS videos https://www.cabidigitallibrary.org/doi/book/10.1079/9781786391582.0000. Please note: this video demonstrates methods of stunning and includes depictions of animals being killed, which some viewers may find upsetting. The use of certain methods and approaches shown may not be permissible in some countries.

Legislation

In developed countries such as the UK and the EU, robust legislation is available that protects animal welfare during transport[209] and killing.[210] Legal requirements that support implementation and enforcement include the use of CCTV cameras and the presence of veterinary surgeons (Official Veterinarians) and Animal Welfare Officers in slaughterhouses.

Official Veterinarians play a crucial role in verifying slaughterhouse compliance, including reviewing CCTV footage for potential issues. The welfare inspection process starts when animals arrive at the slaughterhouse, focusing on proper unloading, assessing their fitness for transport, recording the number of fatalities during transport and checking for signs of welfare issues that may have originated on the farm. If any issues are identified related to transport or the farm of origin, the appropriate authorities will be notified to conduct further investigation. Included are checks for bruises and wounds. In the EU and the UK, it is a legal requirement to evaluate the results of post-mortem inspections of meat chickens. This helps identify poor welfare conditions on their farm of origin.

There is less regulatory oversight when animals are killed on farms. Local veterinary professionals can assist farm personnel in understanding and improving their skills in on-farm killing practices. They can also encourage the adoption of official licensing systems, which allow for the independent verification of competence in on-farm killing and related skills.

Voluntary standards

Some retailers and independent organisations conduct audits that exceed basic welfare legal requirements. Some are developing welfare standards that cover the entire life of the animal, including transport and killing. For example, in the UK, certain retailers are adopting RSPCA Assured standards for different species and integrating them into their guidelines. Additionally, the Soil Association has specific slaughter standards that must be met to use their organic label.[211]

Bibliography

Appleby MC, Mench JA, Olsson AS, Hughes BO. Animal welfare. 2nd ed. Wallingford (UK): CABI; 2011.

Beaver BV, Höglund DL. Efficient livestock handling: the practical application of animal welfare and behavioural science. London (UK): Academic Press; 2016.

Bowell T, Cowen R, Kemp G. Critical thinking: a concise guide. 5th ed. Abingdon (UK): Routledge; 2020.

Bradshaw JWS, Casey RA, Brown SL. The behaviour of the domestic cat. 2nd ed. Wallingford (UK): CABI; 2012.

Bradshaw J. Cat sense: how the new feline science can make you a better friend to your pet. London (UK): Penguin Books; 2014.

Broom DM. Sentience and animal welfare. Wallingford (UK): CABI; 2014.

Broom DM, Fraser AF. Broom and Fraser's domestic animal behaviour and welfare. 6th ed. Wallingford (UK): CABI; 2021.

Brown MT. The ethical process: approach to disagreements and controversial issues. 3rd ed. Upper Saddle River (NJ): Prentice Hall; 2003.

Dawkins MS. Through our eyes only? The search for animal consciousness. Oxford (UK): Oxford University Press; 1998.

Dawkins MS. Why animals matter: animal consciousness, animal welfare and human wellbeing. Oxford (UK): Oxford University Press; 2016.

Ferguson D, Doyle R, Lee C, Fisher A, editors. Advances in sheep welfare. 2nd ed. Duxford (UK): Woodhead Publishing; 2026.

EFSA Panel on Animal Health and Welfare (AHAW). Statement on the use of animal-based measures to assess the welfare of animals. EFSA J. 2012;10(6):2767.

EFSA Panel on Animal Health and Welfare (AHAW). Scientific opinion on the welfare risks related to the farming of sheep for wool, meat and milk production. EFSA J. 2014;12(12):3933.

EFSA Panel on Animal Health and Welfare (AHAW). Scientific Opinion on the slaughter of animals: poultry. EFSA J. 2019;17(11):5849

EFSA Panel on Animal Health and Welfare (AHAW). Scientific Opinion on the welfare of rabbits farmed in different production systems. EFSA J. 2020;18(1):5944

EFSA Panel on Animal Health and Welfare (AHAW). Scientific Opinion on the welfare of pigs at slaughter. EFSA J. 2020;18(6):6148

EFSA Panel on Animal Health and Welfare (AHAW). Scientific Opinion on the welfare of domestic birds and rabbits transported in containers. EFSA J. 2022;20(9):7441

EFSA Panel on Animal Health and Welfare (AHAW). Scientific Opinion on the welfare of pigs during transport. EFSA J. 2022;20(9):7445

EFSA Panel on Animal Health and Welfare (AHAW). Scientific Opinion on the welfare of cattle during transport. EFSA J. 2022;20(9):7442

EFSA Panel on Animal Health and Welfare (AHAW). Scientific Opinion on the welfare of small ruminants during transport. EFSA J. 2022;20(9):7404

EFSA Panel on Animal Health and Welfare (AHAW). Scientific Opinion on the welfare of pigs on farm. EFSA J. 2022;20(8):7421

EFSA Panel on Animal Health and Welfare (AHAW). Scientific Opinion on the welfare of laying hens on farm. EFSA J. 2023;21(2):7789

EFSA Panel on Animal Health and Welfare (AHAW). Scientific Opinion on the welfare of broilers on farm. EFSA J. 2023; 21(2):7788

EFSA Panel on Animal Health and Welfare (AHAW). Scientific Opinion on the welfare of beef cattle. EFSA J. 2025;23(7):e9518

EFSA Panel on Animal Health and Welfare (AHAW). Scientific Opinion on the welfare of dairy cows. EFSA J. 2023;21(5):e07993

Eurogroup for Animals Position Paper. Enhanced animal welfare, Method of Production (MoP+) labelling and sustainability labelling. Eurogroup for Animals; 2022 Nov 21. Available at: https://www.eurogroupforanimals.org/library/enhanced-animal-welfare-method-production-mop-labelling-and-sustainability-labelling

European Commission: Directorate-General for Health and Food Safety, ICF, Maestre, M., Campbell, L., Etienne, J. *et al.*, Study on animal welfare labelling – Final report, Publications Office of the European Union, 2022. Available at: https://data.europa.eu/doi/10.2875/676603

Federation of Veterinarians of Europe (FVE). FVE comments on farmed rabbits [Internet]. Brussels (BE): FVE; 2017.

Flecknell P, Waterman-Pearson A. Pain management in animals. Milton Keynes (UK): Saunders; 2000.

Fraser D. Understanding animal welfare: the science in its cultural context. 2nd ed. (UFAW Animal Welfare Series). Chichester (UK): Wiley-Blackwell; 2023.

Fraser, D., Toward a synthesis of conservation and animal welfare science. Anim. Welf. 2010;19(2): 121–124.

Fraser, D., A 'practical' ethic for animals. J. Agric. Environ. Ethics. 2012;25(5): 721–746

Grandin T, editor. Livestock handling and transport. 5th ed. Wallingford (UK): CABI; 2019.

Grandin T, Cockram M, editors. The slaughter of farmed animals: practical ways of enhancing animal welfare. Wallingford (UK): CABI; 2020.

Grandin T, editor. Improving animal welfare: a practical approach. 3rd ed. Wallingford (UK): CABI; 2021.

Gregory NG. Physiology and behaviour of animal suffering. (UFAW Animal Welfare Series). Oxford (UK): Wiley-Blackwell; 2004.

Harvey, A.M., Ramp, D., Mellor, D.J. Review of the foundational knowledge required for assessing horse welfare. Animals (Basel). 2022;12(23):3385

Hemsworth PH, Coleman GJ. Human–livestock interactions: the stockperson and the productivity and welfare of intensively farmed animals. 2nd ed. Wallingford (UK): CABI; 2010

Hubrecht RC. The welfare of animals used in research: practice and ethics. (UFAW Animal Welfare Series). Chichester (UK): Wiley-Blackwell; 2014.

Humane Slaughter Association Resources. Humane Slaughter Association (HSA). https://www.hsa.org.uk/publications/publications

Klabbers J. International law. 3rd ed. Cambridge: Cambridge University Press; 2021.

Kristiansen TS, Ferraio A, Pavlidis MA, van de Vis H, editors. The welfare of fish. Cham (CH): Springer; 2020.

Lester S, Mercurio B, Davies A. World trade law: texts, materials and commentary. 2nd ed. Oxford (UK): Hart Publishing; 2012.

NC3Rs., Housing and Husbandry: Mouse. https://nc3Rs.org.uk/3rs-resources/housing-and-husbandry-mouse (accessed 12 January 2025)

Nicol CJ. The behavioural biology of chickens. Wallingford (UK): CABI; 2015.

PDSA Animal Wellbeing (PAW) Reports [Internet]. Peoples' Dispensary for Sick Animals. [cited 2025 Apr 22]. Available from: https://pdsa.org.uk/what-we-do/pdsa-animal-wellbeing-report

Radford M. Animal welfare law in Britain: regulation and responsibility. Oxford (UK): Oxford University Press; 2001.

Rollin BE. An introduction to veterinary medical ethics: theory and cases. 2nd ed. Oxford (UK): Blackwell Publishing; 2006.

Serpell JA, editor. The domestic dog: its evolution, behaviour and interactions with people. 2nd ed. Cambridge (UK): Cambridge University Press; 2017.

Voluntary initiative group on equines under the EU Platform on Animal Welfare. Guide to good animal welfare practice for the keeping, care, training and use of horses [Internet]. 2019 [cited 2025 May 11]. Avalable from: https://food.ec.europa.eu/system/files/202207/aw_platform_plat_conc_guide_equidae_en.pdf

Wolfensohn SE, Lloyd M. Handbook of laboratory animal management and welfare. 4th ed. Oxford (UK): Wiley-Blackwell; 2013.

References

1. American Veterinary Medical Association (AVMA), Federation of Veterinarians of Europe (FVE) and Canadian Veterinary Medical Association (CVMA). Joint statement on the roles of veterinarians in promoting animal welfare [Internet]. April 2020 [cited 2025 Apr 14]. Available from: https://fve.org/publications/avma-fve-cvma-joint-statement-on-the-roles-of-veterinarians-in-promoting-animal-welfare-april-2020/
2. Dawkins MS. The science of animal suffering. Ethology. 2008;114:937–45.
3. Harrison R. Animal machines. London: Vincent Stuart Publishers Ltd; 1964.
4. Brambell PFW. Report of the Technical Committee to Enquire into the Welfare of Animals kept under Intensive Livestock Husbandry Systems [Internet]. London; 1965 [cited 2023 Jul 13]. Available from: http://webarchive.nationalarchives.gov.uk/20121010012427
5. Animal Welfare Committee [Internet]. [cited 2024 Nov 6]. Available from: https://www.gov.uk/government/groups/farm-animal/welfare-committee-fawc
6. Webster J. The Five Freedoms: a personal history [video on the Internet]. Sydney (NSW): One Welfare; 2020 [cited 2025 Nov 4]. Available from: https://onewelfare.sydney.edu.au/five-freedoms/
7. Mellor DJ. Updating animal welfare thinking: moving beyond the 'five freedoms' towards 'A life worth living.' Animals (Basel). 2016;6(3):21.
8. Assessment Protocols [Internet]. Welfare Quality Network. 2018 [cited 2025 Apr 15]. Available from: https://www.welfarequalitynetwork.net/en-us
9. World Organisation for Animal Health. Section 7: Animal Welfare [Internet]. Terrestrial Code Online Access. [cited 2025 Apr 15]. Available from: https://www.woah.org/en/what-we-do/standards/codes-and-manuals/terrestrial-code-onine-access/
10. Broom DM. Broom and Fraser's domestic animal behaviour and welfare. 6th ed. Wallingford: CABI; 2021.
11. David Bayvel, A.C. and Cross N. Animal Welfare: A Complex Domestic and International Public-Policy Issue—Who Are the Key Players? J Vet Med Educ. 2010;37(1):3-12
12. Stolba A, G. M. Wood-Gush D. The behavior of pigs in a semi-natural environment. Anim Prod. 1989;48:419–25.

13. Beaver BV, Höglund DL. Efficient livestock handling: the practical application of animal welfare and behavioural science. London (UK): Academic Press; 2016.
14. Hemsworth PH, Coleman GJ. Human–livestock interactions: the stockperson and the productivity and welfare of intensively farmed animals. 2nd ed. Wallingford (UK): CABI; 2010.
15. Le Neindre P, Bernard E, Boissy A, Boivin X, Calandreau L, Delon N, Deputte B,Desmoulin-Canselier S, Dunier M, Faivre N, Giurfa M, Guichet J-L, Lansade L, Larrère R, Mormède P,Prunet P, Schaal B, Servière J, Terlouw C, 2017. Animal consciousness. EFSA supporting publication 2017:EN-1196. 165 pp. doi:10.2903/sp.efsa.2017.EN-1196
16. EFSA Panel on Animal Health and Welfare (AHAW). Scientific Opinion related to the aspects of the biology and welfare of animals used for experimental and other scientific purposes. EFSA J. 2005;3(12):292.
17. Wickens S. Review of the evidence of sentience in cephalopod molluscs and decapod crustaceans. Anim. Welf. 2022;31(1): 155–6.
18. Mellor DJ, Diesch TJ, Gunn AJ, Bennet L. The importance of 'awareness' for understanding fetal pain. Brain Res Rev. 2005;49(3):455–71.
19. EFSA Panel on Animal Health and Welfare (AHAW). Animal welfare aspects in respect of the slaughter or killing of pregnant livestock animals (cattle, pigs, sheep, goats, horses). EFSA J. 2017;15(5):e04782.
20. Pattinson D, Fitzgerald M. The neurobiology of infant pain: development of excitatory and inhibitory neurotransmission in the spinal dorsal horn. Reg Anesth Pain Med. 2004;29(1):36–44.
21. Farm Animal Welfare Committee (FAWC). Economics and Farm Animal Welfare [Internet]. 2011 Dec [cited 2017 Mar 20]. Available from: https://assets.publishing,service.gov.uk/media/5a7d5b5840f0b60a7f1aa056/FAWC_report_on_economics_and_farm_animal_welfare/
22. AWIN (Animal Welfare Indicators Network) welfare assessment protocols [Internet]. Animal Welfare Science Hub. [cited 2025 Apr 15]. Available from: https://animalwelfarehub.org/awin-protocols/
23. AssureWeL Improving farm animal welfare through welfare outcome assessment [Internet]. [cited 2025 Apr 15]. Available from: www.assurewel.org
24. Special Eurobarometer SP533: Attitudes of Europeans towards animal welfare [Internet]. European Data. Brussels; 2023 [cited 2025 Apr 15]. Available from: https://data.europa.eu/data/datasets/s2996_99_1_sp533_eng?locale=en.
25. McInerney JP. Animal welfare, economics and policy: report on a study undertaken for the Farm & Animal Health Economics Division of Defra. London (UK): Department for Environment, Food and Rural Affairs (DEFRA); 2004 Feb.
26. Department for Environment, Food and Rural Affairs (DEFRA). Policy paper: Animal Health and Welfare Pathway [Internet]. London (UK): DEFRA; 2024 [cited 2025 Apr 15]. Available from: https://www.gov.uk/government/publications/animal-health-and-welfare-pathway/animal-health-and-welfare-pathway
27. Madre M, Campbell L, Etienne J, Cook E, Matulina A. Study on Animal Welfare Labelling: Final Report [Internet]. European Commission: Directorate – General for Health and Food Safety. Publications Office of the European Union; 2022 [cited 2025 Apr 15]. Available from: https://op.europa.eu/en/publications-detail/-/publication/49b6b125-b0a3-11ec-83e1-01aa75ed71a1

28. Business Benchmark for Animal Welfare (BBFAW) [Internet]. [cited 2024 Nov 16]. Available from: https://www.bbfaw.com
29. FAI [Internet]. [cited 2024 Nov 5]. Available from: https://www.faifarms.com
30. Farm Animal Welfare Forum. Labelling food from farm animals: method of production labels for the EU: a paper for consultation with stakeholders. London (UK): Farm Animal Welfare Forum; 2010.
31. RSPCA Assured [Internet]. [cited 2024 Nov 6]. Available from: https://rspcaassured.org.uk/
32. Label Rouge [Internet]. [cited 2024 Nov 7]. Available from: https://www.labelrouge.fr
33. Animal Partnership [Internet]. [cited 2024 Nov 7]. Available from: https://globalanimalpartnership.org./
34. Beter Leven [Internet]. [cited 2024 Nov 7]. Available from: https://beterleven.dierenbescherring.nl/
35. SPCA Certified [Internet]. [cited 2024 Nov 7]. Available from: https://www.spcacertified.nz
36. RSPCA Approved Farming Scheme [Internet]. [cited 2025 Apr 21]. Available from: https:/rspcaapproved.org.au
37. Guidance: Poultry meat marketing standards [Internet]. Department for the Environment, Food and Rural Affairs (DEFRA); 2024 [cited 2024 Nov 21]. Available from: https://www.gov.uk/guidance/poultry-meat-marketing-standards
38. Know Your Labels [Internet]. Compassion in World Farming (CIWF). [cited 2025 Apr 21]. Available from: https://ciwf.org.uk/your-food/know-your-labels/
39. British Veterinary Association (BVA). #Choose Assured UK Farm Assurance Scheme Infographic [Internet]. 2022 [cited 2025 Apr 21]. Available from: https://www.bva.co.uk/media/3767/bva-uk-farm-assurance-schemes-choose-assured-infographic.
40. Article 12 (Marking of Packs), Commission Regulation (EC) No 589/2008 of 23 June 2008 laying down detailed rules for implementing Council Regulation (EC) No 1234/2007 as regards marketing standards for eggs. Off J Eur Union. 2008 Jun 24;L163:6–23.
41. UK Egg Industry Data [Internet]. EggInfo. [cited 2025 May 11]. Available from: https://www.egginfo.co.uk/eggs-facts-and-figures/industry-information/data
42. Bloor C, Allan L. Pain scoring systems in the feline and canine patient. Vet Nurse. 2017;8(5):294–301.
43. Tomacheuski RM, Monteiro BP, Evangelista MC, Luna SPL, Steagall PV. Measurement properties of pain scoring instruments in farm animals: a systematic review using the COSMIN checklist. PLoS One. 2023;18(1):e0280037.
44. World Small Animal Veterinary Association (WSAVA). Pain Guidelines [Internet]. [cited 2024 Dec 22]. Available from: https://wsava.org/global-guidelines/pain-guidelines/
45. VetMetrica HRQL, NEWMETRICA Innovation in quality of life assessment [Internet]. [cited 2025 Sep 14]. Available from: https://www.newmetrica.com/vetmetrica-hrql/
46. Rollin BE. An introduction to veterinary medical ethics: theory and cases. 2nd ed. Oxford (UK): Blackwell Publishing; 2006.
47. Brown MT. The ethical process: approach to disagreements and controversial issues. 3rd ed. Upper Saddle River (NJ): Prentice Hall; 2003.
48. Royal College of Veterinary Surgeons. Code of Professional Conduct for Veterinary Nurses [Internet]. Code of Professional Conduct for Veterinary Surgeons [cited 2024 Nov 7]. Available from: https://www.rcvs.org.uk/veterinary-professionals

49. Noddings N. Caring: a feminine approach to ethics. Berkeley: University of California Press; 1984.
50. North DC. Institutions. J Econ Perspect. 1991;5(1):97–112.
51. Riddle E, MacKay JRD. Social media contexts moderate perceptions of animals. Animals (Basel). 2020;10(5):845.
52. Stumpf A, Herbrandt S, Betting L, Kemper N, Fels M. Societal perception of animal videos on social media – funny content or animal suffering? A survey. Animals (Basel). 2024;14(15):2234.
53. Beauchamps TL, Childress JF. Principles of biomedical ethics. 5th ed. New York: Oxford University Press; 2001.
54. Webster J. Ethical and animal welfare considerations in relation to species selection for animal experimentation. Animals (Basel). 2014;4(4):729–41.
55. Russel WMS, Burch RL. The Principles of Humane Experimental Technique [Internet]. Methuen; 1959 [cited 2024 Nov 5]. Available from: https://nerecopa.no/textbase/the-principles-of-humane-experimental-technique
56. Gascoigne E, Mouland C, Lovatt F. Considering the 3Rs for castration and tail docking in sheep. In Pract. 2021;43(3):152–62.
57. Bowell T, Cavan R, Kemp G. Critical thinking: a concise guide. 5th ed. London and New York: Routledge, Taylor & Francis Group; 2020.
58. International Companion Animal Management (ICAM) Coalition [Internet]. [cited 2025 Apr 21]. Available from: https://www.icam-coalition.org
59. Animal Welfare Act 2006 – Explanatory Notes [Internet]. [cited 2024 Nov 7]. Available from: https://www.legislation.gov.uk/ukpga/2006/45/notes
60. Klabbers J. International law. 3rd ed. Cambridge: Cambridge University Press; 2021.
61. Convention on International Trade in Endangered Species of Wild Fauna and Flora (CITES) [Internet]. Washington (DC): CITES Secretariat; 1973 [cited 2025 Nov 4]. Available from: https://cites.org/sites/default/files/eng/disc/CITES-Convention-EN.pdf
62. Corsetti S, Natoli E, Malandrucco L. The dark side of the moon: a good adoption rate conceals the unsolved ethical problem of never-adopted dogs. Animals (Basel). 2025;15(5):670.
63. The Welfare of Farmed Animals (England) Regulations 2007, SI 2007 No. 2078. London (UK): The Stationery Office; 2007.
64. Broom DM. Broom and Fraser's domestic animal behaviour and welfare. 6th ed. Wallingford: CABI; 2022.
65. The Farming Community Network [Internet]. [cited 2025 Apr 22]. Available from: https://fcn.org.uk
66. PDSA Animal Wellbeing (PAW) Reports [Internet]. Peoples' Dispensary for Sick Animals. [cited 2025 Apr 22]. Available from: https://pdsa.org.uk/what-we-do/pdsa-animal-wellbeing-report
67. Article XX: General Exceptions [Internet]. The General Agreement on Tariffs and Trade (GATT). World Trade Organisation; 1947 [cited 2024 Nov 25]. Available from: https://www.wto.org/english/docs_e/legal_e/gatt47_02_e.htm#articleXX.
68. Turtle Excluder Devices [Internet]. NOAA Fisheries. [cited 2025 Apr 22]. Available from: https://www.fisheries.noaa.gov/southeast/bycatch/turtle-excluder-devices
69. Dispute Settlement DS58. United States – Import Prohibitions of certain Shrimp and Shrimp Products [Internet]. World Trade Organisation. [cited 2025 Apr 22]. Available from: https://www.wto.org/english/tratop_e/dispu_e/cases_e/ds58_e.htm.

70. Frequent Questions: Dolphin-Safe, Marine Mammal Protection. [Internet]. NOAA Fisheries; [cited 2025 Apr 22]. Available from: https://www.fisheries.noaa.gov/marine-mammal-protection/frequent-questions-dolphin-safe.
71. World Trade Organization (WTO). Environment: disputes 4 – Mexico etc. versus United States: 'tuna–dolphin' [Internet]. Geneva (CH): WTO; [cited 2025 Apr 22]. Available from: https://www.wto.org/english/tratop_e/envir_e/edis04_e.htm.
72. European Union. Consolidated version of the Treaty on the Functioning of the European Union: Articles 15, 43, 115 and 192. Off J Eur Union. 2016 Jun 7;C202:47.
73. Bradford A. The Brussels effect: how the European Union rules the world. Oxford: Oxford University Press; 2020.
74. European Union. Consolidated version of the Treaty on the Functioning of the European Union: Article 13 [Internet]. EUR-Lex. 2016 [cited 2025 Apr 22]. Available from: https://eur-lex.europa.eu/eli/treaty/tfeu_2016/art_13/oj/eng
75. EFSA Panel on Animal Health and Welfare (AHAW). Scientific opinion on the welfare of pigs on farm. EFSA J. 2022;20(8):7421.
76. EFSA Panel on Animal Health and Welfare (AHAW). Scientific opinion on animal health and welfare in fattening pigs in relation to housing and husbandry. EFSA J. 2007;5(10):611.
77. EFSA Panel on Animal Health and Welfare (AHAW). Scientific opinion on the animal health and welfare aspects of different housing and husbandry systems for adult breeding boars, pregnant, farrowing sows and unweaned piglets. EFSA J. 2007;5(10):572.
78. European Union. Council Directive 2008/120/EC of 18 December 2008 laying down minimum standards for the protection of pigs. Off J Eur Union. 2009 Feb 18;L47:5–13.
79. European Commission. Eurobarometer: public opinion in the European Union [Internet]. Brussels (BE): European Commission; [cited 2025 Apr 23]. Available from: https://europa.eu/eurobarometer/screen/home
80. The Council of Europe [Internet]. [cited 2025 Jul 27]. Available from: https://www.coe.int/en/web/portal
81. Radford M. The First Legislation. In: Animal welfare law in Britain. Oxford: Oxford University Press; 2001. p. 33–59.
82. United Kingdom Parliament. Statement by Mr Hoy during debate on the Protection of Animals Act 1911, 18 March 1964. Hansard (House of Commons Debates). London (UK): UK Parliament; 1964.
83. Kim SW, Gormley A, Jang KB, Duarte ME. Current status of global pig production: an overview and research trends. Anim Biosci. 2024;37(4):719–29.
84. Federation of Veterinarians of Europe (FVE). FVE position paper on moving towards more welfare-friendly farrowing systems [Internet]. 2021 [cited 2025 Apr 23]. Available from: https://fve.org/csm/wp-content/uploads/FVE-position-paper-on-moving-towards-more-welfare-friendly-farrowing-systems-adopted.pdf.
85. Free Farrowing [Internet]. [cited 2025 Apr 23]. Available from: https://www.freefarrowing.org.
86. FareWellDock [Internet]. [cited 2025 Apr 23]. Available from: https://farewelldock.eu
87. EURCAW-Pigs [Internet]. Brussels (Belgium): EU Reference Centre for Animal Welfare for Pigs [cited 2025 Apr 23]. Available from: https://eurcaw-pigs.eu/
88. Food and Agriculture Organization of the United Nations (FAO), Statistics Division. FAOSTAT [Internet]. Rome (IT): FAO; 2023 [cited 2024 Dec 10]. Available from: https://www.fao.org/faostat/en/#home

89. Department for Environment, Food and Rural Affairs (DEFRA). Quarterly UK statistics about eggs – statistics notice (data to Q3 2024) [Internet]. London (UK): DEFRA; 2024 [cited 2025 Apr 23]. Available from: https://www.gov.uk/government/statistics/egg-statistics/quarterly-uk-statistics-about-eggs-statistics-notice-data-to-june-2024
90. Nicol CJ. Behavioural biology of chickens. Wallingford (UK): CABI; 2015.
91. Laying Hen Welfare Forum [Internet]. [cited 2024 Dec 11]. Available from: https://lhwf.co.uk.
92. The Better Chicken Commitment [Internet]. [cited 2024 Dec 10]. Available from: https://betterchickencommitment.com.
93. European Union. Council Directive 1999/74/EC of 19 July 1999 laying down minimum standards for the protection of laying hens. Off J Eur Union. 1999 Aug 3;L203:53–7.
94. European Union. Council Directive 2007/43/EC of 28 June 2007 laying down minimum rules for the protection of chickens kept for meat production. Off J Eur Union. 2007 Jul 12;L182:19–28.
95. Food Animal Initiative (FAI) Farms. Sheep lameness 5 point plan [Internet]. Oxford (UK): FAI Farms; [cited 2025 May 11]. Available from: https://www.faifarms.com/lameness5pointplan/
96. Royal Society for the Prevention of Cruelty to Animals (RSPCA). RSPCA welfare standards: sheep [Internet]. Horsham (UK): RSPCA; 2023 [cited 2025 May 11]. Available from: https://business.rspcaassured.org.uk/media/atiluphr/sheep_welfare_standards_2023.pdf
97. EFSA Panel on Animal Health and Welfare (AHAW). Scientific Opinion on the welfare risks related to the farming of sheep for wool, meat and milk production. EFSA J. 2014;12(12):3933.
98. British Veterinary Association (BVA). Response to International Trade Committee inquiry on UK trade negotiations: agreement with Australia [Internet]. London (UK): BVA; 2021 [cited 2025 May 11]. Available from: https://www.bva.co.uk/media/4465/bva-consultation-response-international-trade-committee-australia-final.pdf
99. Rostami A, Riahi SM, Hofmann A, Ma G, Wang T, Behniafar H, et al. Global prevalence of Toxocara infection in dogs. Adv Parasitol. 2020;109:561–83.
100. Worldostats. Dog population by country: 2025 (updated list) [Internet]:[cited 2025 May 12]. Available from: https://worldostats.com/country-stats/dog-population-by-country/
101. World Health Organization (WHO). One Health [Internet]. Geneva (CH): WHO; 2017 [cited 2024 Dec 28]. Available from: https://www.who.int/news-room/questions-and-answers/item/one-health
102. Roberts C, Blackwell EJ, Roe E, Murrell JC, Mullan S. Awareness and use of canine quality of life assessment tools in UK veterinary practice. Animals (Basel). 2023;13(6):1021.
103. People's Dispensary for Sick Animals (PDSA). PDSA offers PetWise MOT workshops for veterinary profession [Internet]. Vetclick News; 2017 Feb 14 [cited 2025 Nov 4]. Available from: https://www.vetclick.com/news/pdsa-offers-petwise-mot-workshops-for-veterinary-profession-p4523.php
104. Jeanne Marchig International Centre for Animal Welfare Education (JMICAWE). Canine catch, neuter and return (CNR) [Internet]. Edinburgh (UK): University of Edinburgh; 2021 [cited 2025 Apr 24]. Available from: https://media.ed.ac.uk/channel/Canine+Catch+Neuter+Return+%28CNR%29/121598091

105. Barnard S, Pedernera C, Velarde A, Dalla Villa P, Arena L. Shelter Quality: welfare assessment protocol for shelter dogs [Internet]. Teramo (IT): Istituto Zooprofilattico Sperimentale dell'Abruzzo e del Molise "G. Caporale"; 2014 [cited 2025 Apr 24]. Available from: https://izs.it/IZS/Engine/RAServeFile.php/f/pdf_pubblicazioni/ProcolloShelterQuality_EN_2016-DEF.pdf
106. Rooney N, Sargan D. Pedigree dog breeding in the UK: a major welfare concern? [Internet]. Horsham (UK): Royal Society for the Prevention of Cruelty to Animals (RSPCA); 2009 [cited 2024 Dec 29]. Available from: https://www.rspca.org.uk/webContent/staticImages/Downloads/PedigreeDogsReport.pdf
107. Associate Parliamentary Group for Animal Welfare (APGAW). A healthier future for pedigree dogs. London (UK): APGAW; 2009 Nov.
108. Bateson P. Independent inquiry into dog breeding [Internet]. Cambridge (UK): University of Cambridge; 2010 [cited 2025 May 12]. Available from: https://www.dogbreedinginquiry.com
109. Dog Breeding Reform Group (DBRG). Dog Breeding Reform Group [Internet]. [place unknown]: DBRG; [cited 2024 Dec 28]. Available from: https://dbrg.org.uk
110. Universities Federation for Animal Welfare (UFAW). Genetic welfare problems of companion animals [Internet]. Wheathampstead (UK): UFAW [cited 2024 Dec 28]. Available from: https://www.ufaw.org.uk/genetic-welfare-problems-intro/
111. Animal Welfare Foundation (AWF). The Puppy Contract [Internet]. London (UK): AWF; [cited 2025 Apr 24]. Available from: https://www.animalwelfarefoundation.org.uk/education/animal-welfare-advice/puppy-contract
112. British Veterinary Association (BVA). Brachycephalic dogs [Internet]. London (UK): BVA; [cited 2025 Apr 24]. Available from: https://www.bva.co.uk/take-action/our-policies/brachycephalic-dogs/
113. The Welfare of Animals (Licensing of Activities Involving Animals) (England) (Amendment) Regulations 2019, SI 2019/1093. London (UK): The Stationery Office; 2019.
114. Federation of Veterinarians of Europe (FVE), Federation of European Equine Veterinary Associations (FEEVA), Federation of European Companion Animal Veterinary Associations (FECAVA), World Small Animal Veterinary Association (WSAVA). Joint position paper on the animal welfare implications of animal behavioural modification training methods and ability to express species-specific behaviours [Internet]. Brussels (BE): FVE; 2024 [cited 2024 Dec 28]. Available from: https://fve.org/publications/fve-feeva-fecava-and-wsava-joint-position-paper-on-the-behaviour-and-training-of-animals/
115. Companion Animal Welfare Council (CAWC). Report on companion animal welfare establishments: sanctuaries, shelters and re-homing centres [Internet]. London (UK): CAWC; 2004 [cited 2025 Apr 24]. Available from: https://edepot.wur.nl/1560
116. Carter E. CAWC report on the rescue and re-homing of companion animals. Anim Welf. 2011;20(4):652.
117. Worldostats. Cat population by country: 2025 (updated list) [Internet]: Worldostats; [cited 2025 May 12]. Available from: https://worldostats.com/country-stats/cat-population-by-country/
118. People's Dispensary for Sick Animals (PDSA). PAW (PDSA Animal Wellbeing) Report 2022 [Internet]. London (UK): PDSA; 2022 [cited 2025 Jan 11]. Available from: https://www.pdsa.org.uk/media/12965/pdsa-paw-report

119. People's Dispensary for Sick Animals (PDSA). PAW (PDSA Animal Wellbeing) Report 2023 [Internet]. London (UK): PDSA; 2023 [cited 2024 Nov 16]. Available from: https://www.pdsa.org.uk/what-we-do/pdsa-animal-wellbeing-report/paw-report-2023
120. Weiss S, Slater M, Lord L. Frequency of Lost Dogs and Cats in the United States and Methods Used to Locate Them. Animals (Basel). 2012;2(2):301–15.
121. Rochlitz I. A review of the housing requirements of domestic cats (*Felis silvestris catus*) kept in the home. Appl Anim Behav Sci. 2005;93(1–2):97–109.
122. Vojtkovská V, Voslářová E, Večerek V. Methods of assessment of the welfare of shelter cats: a review. Animals (Basel). 2020;10(3):559.
123. Taffin ERL, Paepe D, Campos M, Duchateau L, Goris N, De Roover K, et al. Evaluation of a modified Karnofsky score to assess physical and psychological wellbeing of cats in a hospital setting. J Feline Med Surg. 2016;18(11):913–20.
124. Freeman LM, Rush JE, Oyama MA, MacDonald KA, Cunningham SM, Bulmer BJ, et al. Development and initial validation of the Cat Health and Wellbeing (CHEW) questionnaire. J Feline Med Surg. 2016;18(9):689–701.
125. Feline behaviour and problem behaviours [Internet]. J Feline Med Surg. Special issue. [cited 2024 Nov 16]. Available from: https://journals.sagepub.com/jfm/behaviour
126. Bradshaw J. Cat sense: how the new feline science can make you a better friend to your pet. London (UK): Penguin Books; 2014.
127. Federation of Veterinarians of Europe (FVE). FVE comments on farmed rabbits [Internet]. Brussels (BE): FVE; 2017.
128. EFSA Panel on Animal Health and Welfare (AHAW). Welfare of rabbits on farms. EFSA J. 2020;18(1):e05944.
129. Seaman SC, Waran NK, Mason GJ, D'Eath RB. Animal economics: assessing the motivation of female laboratory rabbits to reach a platform, social contact and food. Anim Behav. 2008;75(1):31–42.
130. Bradbury AG, Dickens GJ. Handling pet rabbits: a review. J Small Anim Pract. 2016;57(10):503–9.
131. Hansen LT, Berthelsen H. The effect of environmental enrichment on the behaviour of caged rabbits (*Oryctolagus cuniculus*). Appl Anim Behav Sci. 2000;68(2):163–78.
132. Rabbit Welfare Association and Fund (RWAF) [Internet]. [cited 2024 Nov 16]. Available from: https://rabbitwelfare.co.uk
133. Animal Behaviour and Training Council (ABTC) [Internet]. [cited 2024 Nov 16]. Available from: https://abtc.org.uk
134. French Agency for Food, Environmental and Occupational Health & Safety (ANSES). On-farm rabbit welfare indicators [Internet]. Maisons-Alfort (FR): ANSES; 2022 [cited 2025 Jan 2]. Available from: https://sitesv2.anses.fr/en/system/files/VF%20List%20of%20welfare%20indicators%20rabbits.pdf
135. British Veterinary Association (BVA). Housing pet rabbits in pairs or compatible groups [Internet]. London (UK): BVA; [cited 2025 May 11]. Available from: https://www.bva.co.uk/take-action/our-policies/housing-pet-rabbits-in-pairs-or-compatible-groups/
136. National Centre for the Replacement, Refinement and Reduction of Animals in Research (NC3Rs). Housing and husbandry: rabbits [Internet]. London (UK): NC3Rs; [cited 2024 Nov 16]. Available from: https://nc3rs.org.uk/3rs-resources/housing-and-husbandry-rabbit

137. European Union Reference Centre for Animal Welfare Poultry and Small Farmed Animals (EURCAW-Poultry-SFA) [Internet]. Wageningen (NL): EURCAW-Poultry-SFA; [cited 2025 May 11]. Available from: https://eurcaw-poultry-sfa.eu
138. Benedetti B, Magrin L, Ferri N, et al. A review of the literature on horse welfare: foundations and future perspectives. Ital J Anim Sci. 2023;22(1):1095–1109.
139. Harvey AM, Ramp D, Mellor DJ. The foundations of equine welfare: a review of equine behaviour and the human–horse relationship. Animals (Basel). 2022;12(15):1889.
140. Brooke. Standardised Equine-Based Welfare Assessment Tool (SEBWAT): overview [Internet]. London (UK): The Brooke; [cited 2025 May 11]. Available from: https://www.thebrooke.org/sites/default/files/Professionals/SEBWAT_0.pdf
141. Department for Environment, Food and Rural Affairs (DEFRA). Code of practice for the welfare of horses, ponies, donkeys and their hybrids [Internet]. London (UK): DEFRA; 2017 [cited 2019 Jan 10]. Available from: https://assets.publishing.service.gov.uk/government/uploads/system/uploads/attachment_data/file/700200/horses-welfare-codes-of-practice-april2018.pdf
142. Voluntary initiative group on equines under the EU Platform on Animal Welfare. Guide to good animal welfare practice for the keeping, care, training and use of horses [Internet]. 2019 [cited 2025 May 11]. Available from: https://food.ec.europa.eu/system/files/2022-07/aw_platform_plat_conc_guide_equidae_en.pdf
143. World Horse Welfare, Eurogroup for Animals. Removing the blinkers: the health and welfare of European equidae in 2015 [Internet]. Brussels (BE): Eurogroup for Animals; 2015 [cited 2025 May 11]. Available from: https://eu.worldhorsewelfare.org/what-we-do/eu-reports
144. Nielsen SS, Alvarez J, Bicout DJ, Calistri P, Depner K, Drewe JA, et al. Welfare of equidae during transport. EFSA J. 2022;20(9):7540.
145. Nielsen SS, Alvarez J, Bicout DJ, Calistri P, Depner K, Drewe JA, et al. Welfare of horses at slaughter. EFSA J. 2025;23(1):8562.
146. European Union Reference Centre for Animal Welfare – Ruminants and Equines (EURCAW-Ruminants & Equines) [Internet]. Wageningen (NL): EURCAW; [cited 2025 May 11]. Available from: https://www.eurcaw-ruminants-equines.eu
147. European Food Safety Authority (EFSA) Panel on Animal Health and Welfare (AHAW). Scientific opinion on the welfare of fish and the concept of sentience. EFSA J. 2009;954:1–27.
148. Utne-Palm AC, Smith A. Fish as Laboratory Animals. In: Kristiansen TS, Ferno A, Pavlidis MA, van de Vis H, editors. The Welfare of Fish. Cham (CH): Springer; 2020. p. 375–400.
149. Aquatic Life Institute. Welfare in marine capture fisheries: policy recommendations [Internet]. New York (NY): Aquatic Life Institute; 2024 Mar [cited 2025 May 11]. Available from: https://ali.fish/publications
150. Kleiber A, Stomp M, Rouby M, Ferreira VHB, Bégout ML, Benhaïm D, et al. Cognitive enrichment to increase fish welfare in aquaculture: a review. Aquaculture. 2023;575:739806.
151. Nofima. Download the Fishwell handbooks [Internet]. Tromsø (NO): Nofima; [cited 2025 May 11]. Available from: https://nofima.com/press-release/download-the-fishwell-handbooks
152. Asociación Empresarial de Acuicultura de España (APROMAR). A guide to fish welfare in Spanish aquaculture – volume 1: concepts and generalities [Internet].

Cádiz (ES): APROMAR; 2022 [cited 2025 May 11]. Available from: https://apromar.es/wp-content/uploads/2022/11/APROMAR-2022_Fish-Welfare-in-Spanish-aquaculture.pdf

153. Food Animal Initiative (FAI) Farms. Driving aquaculture welfare improvements [Internet]. Oxford (UK): FAI Farms; [cited 2025 May 10]. Available from: https://www.faifarms.com/driving-improvements-for-aquaculture-welfare/
154. European Union Reference Centre for Animal Welfare – Aquatic Animals (EURCAW-Aqua). [Internet]. Wageningen (NL): EURCAW; [cited 2025 Feb 17]. Available from: https://www.eurcaw-aqua.eu
155. Stevens CH, Croft DP, Paull GC, Tyler CR. Stress and welfare in ornamental fishes: what can be learned from aquaculture. J Fish Biol. 2017;91(2):409–28.
156. Rozenbaum M. FBI: the bureau for fish welfare [Internet]. London (UK): Understanding Animal Research; 2020 [cited 2025 May 10]. Available from: https://www.understandinganimalresearch.org.uk/news/fbi-the-bureau-for-fish-welfare
157. Royal Society for the Prevention of Cruelty to Animals (RSPCA). RSPCA welfare standards for farmed rainbow trout [Internet]. Horsham (UK): RSPCA; 2020 [cited 2025 May 10]. Available from: https://business.rspcaassured.org.uk/media/mimbtfua/trout-standards-rspca-2020.pdf
158. Royal Society for the Prevention of Cruelty to Animals (RSPCA). RSPCA welfare standards for farmed Atlantic salmon [Internet]. Horsham (UK): RSPCA; 2024 [cited 2025 May 11]. Available from: https://business.rspcaassured.org.uk/media/mzjhvulj/rspca-welfare-standards-salmon-2024.pdf
159. Carballeira Brana CB, Cerbule K, Senff P, Stolz IK. Towards environmental sustainability in marine finfish aquaculture. Front Mar Sci. 2021;8:656843.
160. Farmers and animal advocates discuss fish welfare in Holyrood [Internet]. The Fish Site; 2025 [cited 2025 May 11]. Available from: https://www.thefishsite.com/articles/farmers-and-animal-advocates-discuss-fish-welfare-in-holyrood
161. Animal Welfare Committee (AWC). Update to the 2014 FAWC opinion on the welfare of farmed fish at the time of killing [Internet]. London (UK): Department for Environment, Food and Rural Affairs (DEFRA); 2023 [cited 2025 May 11]. Available from: https://www.gov.uk/government/groups/animal-welfare-committee-awc
162. European Parliament. Animal welfare of farmed fish: study requested by the Committee on Fisheries (PECH Committee) [Internet]. Brussels (BE): European Parliament; 2023 [cited 2025 May 11]. Report No.: PE 747.257.
163. Bar-On YM, Phillips R, Milo R. The biomass distribution on Earth. Proc Natl Acad Sci U S A. 2018;115(25):6506–11. Supplementary information: Appendix 1 [Internet]. Available from: https://www.pnas.org/doi/suppl/10.1073/pnas.1711842115
164. Tomasik B. How many wild animals are there? [Internet]. Reducing Suffering; 2019 [cited 2025 Jan 11]. Available from: https://reducing-suffering.org
165. Wild Animal Welfare Committee (WAWC). Live trapping of terrestrial wild mammals and wild birds for lethal control in the United Kingdom [Internet]. London (UK): WAWC; [cited 2025 May 11]. Available from: https://www.wawcommittee.org/resources
166. Goodnature Traps UK. Lethal trapping of terrestrial mammals [Internet]. Edinburgh (UK): Goodnature Traps [cited 2025 May 11]. Available from: https://goodnaturetraps.co.uk
167. Simmons A. Why are different animal welfare standards applied to different animals in different settings? In: Wild Animal Welfare Committee (WAWC) Conference 2021; 2021 Nov; London (UK).

168. Fraser D. Toward a synthesis of conservation and animal welfare science. Anim Welf. 2010;19(2):121–4.
169. Mullineaux E. Veterinary treatment and rehabilitation of indigenous wildlife. J Small Anim Pract. 2014;55(6):293–300.
170. Sprogis KR, Raudino HC, Rankin R, MacLeod CD, Bejder L. Home range size of Indo-Pacific bottlenose dolphins (*Tursiops aduncus*) in a coastal and estuarine system is habitat- and sex-specific. Mar Mamm Sci. 2016;32(1):287–308.
171. Clegg ILK, Borger-Turner JL, Eskelinen HC. C-Well: the development of a welfare assessment index for captive dolphins (Tursiops truncatus). Anim Welf. 2015;24(3):267–82.
172. Nutrition Advisory Group to the Association of Zoos and Aquariums. Body condition scoring resource centre [Internet]. 2024 [cited 2025 Jan 18]. Available from: https://nagonline.net/3877/body-condition-scoring/
173. Dubois S, Fenwick N, Ryan EA, Baker SE, Beausoleil NJ, Carter S, et al. International consensus principles for ethical wildlife control. Conserv Biol. 2017;31(4):753–60.
174. Wild Animal Welfare Committee (WAWC). Wild Animal Welfare Committee [Internet]. [cited 2025 May 11]. Available from: https://www.wawcommittee.org
175. AnimalKind. BC SPCA accredited wildlife and rodent control standards [Internet]. Version 1.8. Vancouver (CA): BC SPCA; 2024 [cited 2025 Apr 24]. Available from: https://animalkind.ca/wp-content/uploads/AnimalKind-Wildlife-Control-Standards.pdf
176. British Veterinary Association (BVA). Green Team Vet [Internet]. London (UK): BVA [cited 2025 May 11]. Available from: https://www.bva.co.uk/take-action/green-team-vet/
177. Saferoad [Internet]. [cited 2025 May 11]. Available from: https://saferoad-cedr.org/en/saferoad.htm
178. British Veterinary Zoological Society (BVZS). Good practice guidelines for wildlife rehabilitation centres [Internet]. London (UK): BVZS; 2016 [cited 2025 May 11]. Available from: https://www.rcvs.org.uk/document-library/bvzs-good-practice-guidelines-for-wildlife-rehabilitation
179. Association of Zoo and Exotic Veterinary Nurses (AZEVN) [Internet]. [cited 2025 May 11]. Available from: https://azevn.org
180. Wildlife Disease Association [Internet]. [cited 2025 May 11]. Available from: https://wildlifedisease.org
181. Department for Environment, Food and Rural Affairs (DEFRA). Secretary of State's standards of modern zoo practice [Internet]. London (UK): DEFRA; 2012 [cited 2025 May 11]. Available from: https://www.gov.uk/government/publications/standards-of-modern-zoo-practice
182. European Association of Zoos and Aquaria (EAZA). Positive animal welfare [Internet]. Amsterdam (NL): EAZA [cited 2025 May 11]. Available from: https://www.eaza.net/welfare
183. Home Office. Statistics of scientific procedures on living animals, Great Britain 2023 [Internet]. London (UK): Home Office; 2024 [cited 2024 Nov 5]. Available from: https://www.gov.uk/government/statistics/statistics-of-scientific-procedures-on-living-animals-great-britain-2023
184. European Commission. Summary report on the statistics on the use of animals for scientific purposes in the Member States of the European Union and

Norway in 2022 [Internet]. Brussels (BE): European Commission; 2024 [cited 2024 Nov 4]. Available from: https://data.consilium.europa.eu/doc/document/ST-12496-2024-INIT/en/pdf

185. Colvin JS, Bohne BA, Harding GW, McEwan DG, Ornitz PM. Skeletal overgrowth and deafness in mice lacking fibroblastic growth factor receptor 3. Nat Genet. 1996;12(4):390–7.
186. National Centre for the Replacement, Refinement and Reduction of Animals in Research (NC3Rs) [Internet]. [cited 2024 Nov 5]. Available from: https://nc3rs.org.uk
187. Understanding Animal Research [Internet]. [cited 2025 May 10]. Available from: https://www.understandinganimalresearch.org.uk
188. American College of Laboratory Animal Medicine (ACLAM). Position statement on adequate veterinary care. J Am Assoc Lab Anim Sci. 2016;55(6):826–8.
189. FELASA/ECLAM/ESLAV Joint Working Group on Veterinary Care. Guidelines for the veterinary care of laboratory animals. Lab Anim. 2008;42(1):1–11.
190. Laboratory Animal Veterinary Association (LAVA). Careers [Internet]. [cited 2025 Feb 17]. Available from: https://www.lava.uk.net/careers
191. Animals (Scientific Procedures) Act 1986, c.14. London (UK): The Stationery Office; 1986.
192. European Commission. EU Reference Laboratory for Alternatives to Animal Testing (EURL ECVAM) [Internet]. Ispra (IT): Joint Research Committee; [cited 2025 May 11]. Available from: https://joint-research-centre.eu.europa.eu/projects-and-activities/reference-and-measurement/european-union-reference-laboratory-alternatives-animal-testing-eurl-ecvam
193. AAALAC International [Internet]. [cited 2025 May 11]. Available from: https://www.aaalac.org
194. Beaver BV, Höglund DL. Efficient livestock handling: the practical application of animal welfare and behavioural science. London (UK): Academic Press; 2016.
195. The Welfare of Animals at the Time of Killing (England) Regulations 2015, SI 2015/1782. London (UK): The Stationery Office; 2015.
196. Council Regulation (EC) No 1099/2009 of 24 September 2009 on the protection of animals at the time of killing. Off J Eur Union. 2009 Nov 18;L 303:1–30.
197. The Welfare of Animals at the Time of Killing (Wales) Regulations 2015, SI 2015/1781 (W.257). Cardiff (UK): The Stationery Office; 2015.
198. Humane Slaughter Association. Online Guides [Internet]. Wheathamlstead (UK): HAS; [cited 2025 Mar 25]. Available from: https://www.hsa.org.uk/publications/on-line-guides
199. EFSA Panel on Animal Health and Welfare (AHAW). Scientific Opinion on the welfare of small ruminants during transport. EFSA J. 2022;20(9):e7456.
200. Animal Welfare Committee (AWC). Advice on preslaughter handling of horses [Internet]. London (UK): Department for Environment, Food and Rural Affairs (DEFRA); 2024 [cited 2025 May 10]. Available from: https://www.gov.uk/government/publications/awc-advice-on-preslaughter-handling-of-horses
201. Grandin T. Livestock behaviour, design of facilities and humane slaughter [Internet]. Fort Collins (CO): Colorado State University; [cited 2025 Apr 24]. Available from: https://www.grandin.com
202. Grandin T. Recommended animal handling and audit guide: a systematic approach to animal welfare [Internet]. Washington (DC): North American Meat Institute; 2021 [cited 2025 Apr 24]. Available from: https://www.meatinstitute.org/sites/default/files/original%20documents/Animal_Handling_Guide_English.pdf

203. European Commission. 2018 factsheets: animal welfare at slaughter and stunning [Internet]. Brussels (BE): European Commission; 2018 [cited 2025 Mar 26]. Available from: https://food.ec.europa.eu/animals/animal-welfare/slaughter-stunning/2018-factsheets_en
204. European Commission. Animal transport guides [Internet]. Brussels (BE): European Commission [cited 2025 May 11]. Available from: https://food.ec.europa/animals/animal-welfare/eu-animal-welfare-legislation/animal-welfare-during-transport/animal-transport-guides_en
205. Animal Welfare Committee (AWC). Advice to government: welfare at killing [Internet]. London (UK): DEFRA; [cited 2025 Mar 25]. Available from: https://www.gov.uk/government/collections/awc-advice-to-government
206. European Food Safety Authority (EFSA). Animal Welfare at Slaughter [Internet]. Parma (IT): EFSA [cited 2025 Mar 25]. Available from: https://www.efsa.europa.eu/en/topics/topic/animal-welfare-slaughter
207. British Veterinary Association (BVA). Welfare of animals at slaughter [Internet]. London (UK): BVA [cited 2025 May 11]. Available from: https://www.bva.co.uk/take-action/our-policies/welfare-of-animal-at-slaughter/
208. American Veterinary Medical Association (AVMA). AVMA guidelines for the euthanasia of animals [Internet]. Schaumburg (IL): AVMA; 2020 [cited 2025 May 11]. Available from: https://www.avma.org/resources-tools/avma-policies/avma-guidelines-euthanasia-animals
209. European Commission. Protection of animals during transport [Internet]. Brussels (BE): European Commission; [cited 2025 Apr 24]. Available from: https://food.ec.europa.eu/animals/animal-welfare/eu-animal-welfare-legislation/animal-welfare-during-transport_en
210. European Commission. Slaughter and stunning [Internet]. Brussels (BE): European Commission; [cited 2025 Apr 24]. Available from: https://food.ec.europa.eu/animals/animal-welfare/eu-animal-welfare-legislation/slaughter-stunning_en
211. Soil Association. Soil Association organic standards for Great Britain: abattoir and slaughtering [Internet]. Version 1.5. Bristol(UK): Soil Association; 2024 [cited 2025 May 10]. Available from: https://soilassociation.org/media/25984/sa-gb-abattoir-slaughter.pdf

Index

Printed and bound by CPI Group (UK) Ltd, Croydon, CR0 4YY

13/09/2026

14961107-0002